T0239421

Japan Nutrition

Teiji Nakamura

Japan Nutrition

 Springer

Teiji Nakamura
Kanagawa University of Human Services
Yokosuka, Kanagawa, Japan

ISBN 978-981-16-6318-5   ISBN 978-981-16-6316-1   (eBook)
https://doi.org/10.1007/978-981-16-6316-1

Translation from the Japanese language edition: Clinical nutritionist: "Japan· Nutrition" unraveled by Teiji Nakamura-For the past, present, and future of Japanese nutrition by Teiji Nakamura, © Teiji Nakamura 2020. Published by Daiichi Shuppan Co. Ltd.. All Rights Reserved.

This Springer imprint is published by the registered company Springer Nature Singapore Pte Ltd.
The registered company address is: 152 Beach Road, #21-01/04 Gateway East, Singapore 189721, Singapore

*Words of commendation from the former
Prime Minister of Japan Junichiro Koizumi*
***"Health comes from the right diet.
This book teaches that nutrition brings
health and well-being to people."***

Introduction

In 2019, the Japanese imperial era changed from Heisei (平成) to Reiwa (令和). It was a big milestone in Japanese history. The ceremonies of succession to the throne were held mainly by the imperial family and the government, and I was able to attend these ceremonies: on February 24, 2019, "The Ceremony for the 30th Anniversary of His Majesty The Emperor's Reign" at the National Theater; October 22, "The Ceremony at the Immortal Religious Hall" held at the Imperial Palace; and November14 and15 "Grand Palace Ceremony."

Together with the legislative, administrative, and judicial officials from the government, and representatives of their respective fields such as entertainers, athletes, cultural figures, scientists, and other people who have had outstanding achievements and have contributed to society, I was able to attend as a representative of nutrition.

About 150 years after the Meiji Restoration, when nutrition was introduced, and about 100 years after the research, education, and practice of nutrition began, "nutrition" was recognized as an area that contributed to the nation at such a turning point. I personally felt the great significance of the moment.

The moment I stepped into the palace for the first time, walked down the red carpet from the entrance, and climbed the steps of the "Houmeiden State Banquet Hall (豊明殿)" step by step, I was proud and firm. "Nutrition" is the source of life, and many nutrition specialists and dietitians are important human resources who support health, medical care, and welfare, but they are simply people who do not receive any special attention from society. It was also the moment when I thought that my efforts of the years I had been pursuing health and happiness were rewarded.

The "Tokyo Nutrition for Growth Summit 2021" was held in the year when the "Tokyo Olympics and Paralympics" were held. It's time for nutrition to appear on the front stage. At this important moment, I thought that I had to understand "nutrition" correctly and sort out its meaning, role, mission, history, and direction for the future. Fortunately, I was able to participate directly in the second half of Japan's nutritional improvement, which has a history of about 100 years. I decided to write this book as one of the embodiments of "Japan Nutrition," which has moved away from nutritional deficiency, turned to the problems of excess, and further extended healthy life expectancy. With this book, many people may understand nutrition as a field, realize its value and appeal, and nutritionists and dietitians feel proud and confident, and I want to make it a book that encourages and enthuses those who are thinking of studying the field. This is what I thought, and so I continued writing. I would like you to read this book.

In the publication process, Dr. Seino Fukue, Director for Nutrition, Health Service Bureau, Ministry of Health, Labour and Welfare, provided valuable guidance. I would also like to express my sincere gratitude to Kanagawa University of Human Services, St. Marianna University School of Medicine, the Japan Dietetic Association, President Shigeru Kurita of Daiichi Publishing, and Professor Emeritus Dr. Andrew R. Durkin of Indiana University.

Yokosuka, Kanagawa, Japan   Teiji Nakamura

Contents

About the Author

Teiji Nakamura graduated from the University of Tokushima School of Medicine in 1972. He worked as a clinical dietitian at Shinjuku Clinic and St. Marianna University School of Medicine Hospital. He was a research student at the Tokyo University School of Medicine since 1978 and received a PhD in 1985. The dissertation topic was "Characteristic of eating behavior in obese people and effectiveness of behavior modification therapy."

In 2003, he became a professor at Kanagawa University of Human Services. After working as chairman and dean, he has been the president from 2011 to the present. Also, he is the president of the Japan Dietetic Association and president of the Japanese Association of Nutritional Science Education. He was chairman of the Organizing Committee of the 15th International Congress of Dietetics in 2008. In 2013, he became the chairman of a study group on "healthy diet," which supports the longevity of Japanese people, by the Ministry of Health, Labor, and Welfare.

He is widely known internationally as a leader in the practice of nutrition.

Chapter 1
Preventing and Treating Disease with Nutrition

Abstract In 1976, the author posted a "Nutrition Consultation Room" sign in a corner of St. Marianna University Hospital. Because non-communicable diseases result from incorrect eating habits, their prevention and treatment first require improved nutrition and diet. At that time, expectations for the development of new drugs were high, and only a few patients visited the counseling room.

However, as he continued to teach enthusiastically, the effects of the diet appeared and the number of patients visiting the room increased. In recognition of the results, the national government has also begun to allow nutritional guidance fees within the public medical insurance system.

The first reason I chose nutrition was because I knew "preventive medicine". Doctors who treat illness are certainly noble professions, but I thought that the profession to create a society where people do not get sick is even more valuable

At university, he began researching vitamin B_6. With a deficiency of only one vitamin, the rat developed fatty liver and oily skin, gradually losing weight and dying. He thought I had to inform many people about the importance of this nutrition.

Today, many people in the world are suffering from hunger and obesity, and solving malnutrition is the greatest challenge for humankind. However, Japan solved the postwar nutritional deficiency, suppressed the westernization of diet after high economic growth, and created a longevity nation. Why did Japan succeed in eradicating the double burden of malnutrition?

The author, who played a central role in the movement, unravels the mystery.

Keywords Preventive medicine · Tokushima University in School of Medicine · Vitamin B_6 deficiency · Nutrition consultation room · Lifestyle-related diseases · Fee for nutritional dietary guidance

1.1   Establishment of the "Nutrition Consultation Room"

1.1.1   Launch of the "Nutrition Consultation Room"

In April 1976, I put up a signboard for a "Nutrition Consultation Room" in a corner of the outpatient department of St. Marianna University Hospital. This was at a time when the concept of adult diseases was changing to that of lifestyle-related diseases. If diseases were caused by inappropriate dietary habits, the first step to prevent or treat them was to improve dietary habits, and that was why we created the Nutrition Consultation Room. This was the first attempt of its kind in Japan. It was not that the hospital administrators and doctors were particularly understanding, but the deputy director of the hospital, who was also the head of the nutrition department at the time, somehow listened to my headstrong opinions and allowed me to borrow a part of his room.

At that time, hypertension, diabetes, and arteriosclerosis, which were on the rise among the Japanese population, were called "adult diseases" because they occurred mainly in adulthood. In other words, they were thought to be diseases that inevitably occur with aging. As blood vessels age, arteriosclerosis tends to occur, and the important issue was to reduce the increase in blood sugar, cholesterol, animal fat, and blood pressure. The development of drugs to reduce these factors was in full swing, and the reason for visiting a university hospital was to receive the latest drugs.

Sure enough, we opened a "Nutrition Consultation Room," but no patients came to the room. Occasionally, I was approached and asked, "Where is the ophthalmology department?" The "Nutrition Consultation Room" had turned into what is now called a "general information room". The location was convenient, and at that time, it was rare to find a "consultation room" in a hospital, so people seemed to think that they could consult us about anything.

One day, a doctor of internal medicine came to the office.

"It's a nice room. What are you planning to do here?"

Indeed, it was a splendid half-room borrowed from the office of the deputy medical director. Partly because I felt I had to be emphatic, I quickly responded, "I'm trying to cure illness with nutrition and diets".

"Boy, if food could cure disease, doctors would have no trouble!"

He left laughing. I will never forget that doctor's words and the look on his face. When I think about it now, this may have been the beginning of the long "journey of nutrition" that followed, with the banner "Let's prevent and treat diseases with the power of nutrition". At the time, such a banner was thought to be nothing more than a fantasy for those who knew nothing about medicine or medical treatment. It was a time when adult diseases such as diabetes, hypertension, and arteriosclerosis were on the rise, and medicine was focused on developing new drugs for them.

It was a lonely beginning. However, there were still patients who visited the "Nutrition Consultation Room". The patients who came to our office were enthusiasts of the so-called "OO health method" and the "XX diet method", who followed a certain food or diet method. They were forced to buy expensive health foods, and some of them suffered from nutritional deficiency due to their extreme diets. I made an effort to listen to all the followers and hear them out. By doing this, I learned a lot.

1.1.2   Extension of Nutrition Counseling

About half a year after the opening, requests for nutritional advice began to come in little by little from cardiologists and doctors treating diabetes who had become close friends of mine. Since the patients who came to the room were very important to us, I spent a good deal of time with each patient.

Because I had to achieve results through diet therapy at all costs, every day, I began to write a "patient diary" of the day's consultations. Why didn't the patient follow the instructions? Was it lack of knowledge? Lack of awareness? Lack of motivation? Was the diet itself wrong?

Every day was a series of reflections and innovations. Doctors could prescribe drugs as a method of treatment, but I could not be effective unless patients believed my stories and improved their diet.

At that time, nutrition in Japan was just emerging from the nutritional deficiency caused by food shortages, and the purpose of nutritional guidance was to provide general nutritional knowledge, adjustment of the nutrients to be taken, and dissemination of appropriate dishes and menus. There were no textbooks or reference books to provide individualized nutritional counseling for patients with different diseases or conditions. Nevertheless, research on adult diseases and nutrition was beginning; new findings were new to patients and physicians were interested in them. However, this method did not motivate patients to come to the clinic more than once, and nutritional consultations ended after the first visit. I read a lot of books on nutritional guidance and counseling in Europe and the United States, and in the midst of groping in the dark, I developed my own method of nutritional guidance that could be applied to individual patients.

Fortunately, the number of patients began to increase thanks to our efforts. Several doctors told me that their patients' blood glucose and blood pressure control improved when they were sent to the nutrition consultation room, and rumors began to circulate among patients.

I have a patient I'll never forget.

Mr. K, a 55-year-old man, had been visiting the hospital for diabetes, hypertension, dyslipidemia, and chronic gastritis for many years and had been taking many sorts of medicine. When he started to take dietary therapy, the effects of the therapy began to appear, his medications could be decreased, and his physical condition improved. Mr. K was the owner of a bathhouse. He sat on the manager's bench and gave health advice to his neighbors, and he knew the shape he was in.

Gradually, the number of patients coming to the hospital increased, but diet therapy was still a minor part of their lives. One day, when I consulted with the hospital director, he advised me to participate actively in mass media. This is what is now called public relations. In 1983, NHK (Japan Broadcasting Corporation), encouraged by Dr. Yoshio Ikeda of Jikei Medical University, produced a program to promote proper diet. I gave regular nutritional counseling to 10 obese middle-aged and elderly patients and followed the patients' daily lives for 6 months in a documentary style. The daily lives of those who had to go on a diet were vividly televised, and the program was well worth watching. Looking back on it now, it was the first diet experience program. All the participants succeeded in losing weight, and their blood sugar, blood pressure, neutral fat, cholesterol, and other test values improved.

These cases were used as a reference, and in 1978, the fee for nutritional dietary guidance by a dietitian was approved as a medical fee item for the first time. At that time, the fee was only 5 points (50 Japanese yen:50 cents US) for nearly 1 h of work and was said to be not even enough to pay for a coffee. RD. Kiku Morikawa, who was the former president of the Dietitians' Association, told me that he was encouraged by Dr. Taro Takemi, a former president of the Medical Association, who told him, "Although the fee is low, it is important to make a start, and it will eventually go up, so please be patient." At present, the first time to visit for a consultation is 260 points(24 USD) for 30 min, and from the second time on, it is 200 points for 15 min.

1.2 Why Did I Choose Nutrition?

One's high school days.

It was when I was in high school that I became interested in nutrition. I was born in a place called Hanaoka in Kudamatsu City (下松市), Yamaguchi Prefecture (山口県), and grew up there until high school. Hanaoka (花岡) is an inn town that developed around a gate in front of Hanaoka Hachiman Shrine (花岡八幡宮), which was built in 709, after a sacred object was brought there from Usa Hachiman Shrine (宇佐八幡宮) in Oita Prefecture (大分県). Our house was about 1300 years old, and we are told that it was transferred with the sacred object, which is so old that it makes us feel faint. (Photo 1.1).

At present, this area is home to "Akaibo (閼伽井) Temple", which has a multi-purpose pagoda that is a national treasure, and "Houshouji (法静寺) Temple", which has recently become famous for a strange festival called "Fox Bride (狐の嫁入り)", and the shrine and temple were my childhood playground.

One day, when I was preparing for a university entrance examination, I received an invitation from the chief priest of "Houshouji Temple" to attend a sermon. A Sermon means a religious talk and is often part of a religious service. Since many of my relatives were doctors, I somehow intended to pursue a career in medicine, so I attended. The lecturer that day was a doctor of psychosomatic medicine at Kyushu University School of Medicine, and I heard the word "preventive medicine" for the

Photo 1.1 Hanaoka, Kudamatsu City, Yamaguchi Prefecture, which is depicted in the picture scroll of an intangible cultural property. (Inside the circle: Nakamura family in 1797)

first time. He said, "Doctors who treat diseases certainly have a noble profession, but what is even greater is doctors who create a society in which people do not become ill."

"I was fascinated by these words."

1.2.1 University Days

Determined to study medicine that would prevent illness, I entered the Department of Nutrition at the University of Tokushima Medical School in 1968. (Photo 1.2) The Department of Nutrition is internationally renowned as a research and educational institution for nutrition associated with a medical school and is said to be the Mecca of clinical nutrition. However, when I entered the Department of Nutrition 50 years ago, it was an obscure university, with a mixed group of students who had lost their dreams of becoming doctors or pharmacists, or who had mistakenly entered the Department of Nutrition instead of the Department of Home Economics, and it was unclear what they would study or what their professions would be. Some of my classmates dreamed of becoming chefs and entered the school mistakenly thinking that a national culinary school had been established. Although they were in medical school, they were not going to become doctors or chefs. Furthermore, during their time at the school, none of the faculty members ever told them about the significance or social role of the profession of dietitian or nutritionist.

Photo 1.2 University of Tokushima Medical School

The Department of Nutrition was established by Dr. Keizo Kodama, who was appointed President of the University of Tokushima after serving as Dean of the Faculty of Medicine at the University of Tokyo, with the aim of establishing a human-centered nutrition science based on medical science, as opposed to the food nutrition science that was being studied mainly in domestic science and agriculture at the time. This was a reform that was ahead of its time, as the university was free from nutritional deficiency caused by food shortages and the adverse effects of the westernization of the diet were beginning to occur. The faculty members who came together were all excellent researchers; however, although they had academic interests, they were not interested in the practice of nutrition.

When I consulted with my professor about wanting to work at a hospital before graduation, he told me, "This university is not the place to prepare such professionals". Indeed, the content of the lectures and the curriculum centered on physiology, biochemistry, and clinical medicine, and the university was a training school for nutrition researchers. There was a large, significant disconnect between the education and research at the university and the training of nutritionists. In class, I listened to lectures on physiology, anatomy, and what was even more difficult, biochemistry, while in the hospital kitchen, I spent all day peeling onions and slicing cabbage. As I looked at the piles of shredded cabbage, I thought that the education and research at the university and the work in the field were so unrelated that education at the Department of Nutrition was somehow out of whack.

When I think about it now, at that time, nutrition itself did not yet have an academic system that could form a single department, its role in society was unclear, and its reputation was low, and each faculty member was simply looking at nutrition from the specialized field in which he had been trained. Because the researchers were working strictly from their individual perspectives, the concept of "nutrition" was immature and incomplete, and the value of contributing this knowledge and skill to society and the nature of professional work were not the subject of discussion. Many of the students were educated without knowing what they would be doing after graduation, so they asked themselves, "What kind of professionals or professions are we? What is Nutrition?".

In 2014, the Faculty of Medicine at the University of Tokushima reorganized its existing Department of Nutrition into the Department of Medical Nutrition, aiming to train researchers and educators who are responsible for basic research and education in nutrition, and, at the same time, to train dietitians who can work in cooperation with doctors in clinical areas.

1.3   A Nutrient Deficiency Experiment

A deficiency of just one nutrient can kill people.

I became a senior at university and began to write my graduation thesis. For my seminar, I joined the "Practical Nutrition Class" because I was admired Professor Toshiro Sato. Professor Sato was the grandson of Dr. Shibasaburo Kitazato (北里柴三郎), the first person to introduce medical statistics to Japan, and was a man with a clear mind who was engaged in international research. In class, I took nutritional statistics but hardly knew what it was about. I joined a laboratory, but I had to travel overseas a lot, so I did not have many opportunities to receive guidance. The only thing I remember vividly is that I went to the public health center to pick up a dog for an experiment and helped with the dissection experiment. Therefore, for my graduation thesis, I started research on vitamin B_6 deficiency under the guidance of then Assistant Professor Michiko Okada.

The laboratory rats were kept on a diet deficient in vitamin B_6 and dissected periodically. When the rats were deficient in the vitamin, they developed fatty liver and oily skin, gradually began to lose weight, and finally died. Why would vitamin B_6, a component related to amino acid transferase, be associated with abnormal lipid metabolism? Why does it accumulate in the liver while body fat decreases in the whole body? Solving this mystery was the theme of the study. One doctor asked us whether such a fatty liver could be produced only by vitamin deficiency. It was said that it is not easy to make healthy rats sick, and they cannot produce fatty liver unless they are given something poisonous or a lot of alcohol.

Deficiency of a single nutrient, a not-so-major vitamin, is enough to make one very sick and eventually kill one. As I stood in front of the emaciated and dying rats in the dimly lit animal room, " I have learned that this is a terrible thing and I have to let a lot of people know about this".

At that time, the problem of postwar undernourishment was being solved by economic development, westernization of the diet, and improvements in food distribution, and there was a strong tendency to believe that nutritional problems could be easily solved if only the economy would develop and people would eat nutritious food. One day, a graduate came to the laboratory and asked me what I was researching. I replied with a joke, "I like the Beatles, so I'm researching B6, or vitamin B_6." He laughed at me and said, "Even if you study such things, no one eats while thinking about nutrition." There was a great dissociation between the study of nutrition and its practice, and nutrition research was thought to be like "sashimi no tsuma" a study that was merely a social decoration. In other words, it was thought that malnutrition could be solved by economic prosperity without the need for nutritional research.

1.3.1 A Fateful Encounter and Employment

During the summer vacation of my fourth year, I went to Tokyo on the advice of my older brother (中村隆征) and met a person who changed my life at a small sushi restaurant (Hana Sushi) in a corner of Kabukicho (歌舞伎町). That person was Dr. Hirohisa Arai (新居裕久), the director of Shinjuku Clinic near Seibu Shinjuku Station. He was the first person in Japan to call for "medical nutrition as the source of diets" and gave lectures while cooking with a pot. Although we had never met before, we talked about nutrition, health, food and cooking, and we agreed that the future of medicine is not drugs but nutrition.

After graduation, I moved to Tokyo to work as Dr. Arai' s assistant. He said, "You don't understand clinical medicine, so I will teach you; I don't understand nutrition, so I want you to teach me." My role was to collect and prepare materials for Dr. Arai's lectures and books, and to check his manuscripts and lectures. I often visited the library of Keio University Hospital near Yotsuya Station. I found it interesting to look up the latest nutritional research and the history of nutrition, and I was able to think about this and that, and the time I spent in the library was truly blissful.

Because the clinic was located in the middle of Kabukicho, which is said to be the most prosperous area in Japan, the clinic hours were from 3:00 pm to 8:00 pm. During that time, I observed medical treatment, observed how to conduct a clinical examination, and gave nutritional counseling to patients with obesity, thinness, and adult -onset diseases. At that time, there was no precedent in the medical field for providing nutritional counseling to individuals, so I was truly groping in the dark. I had to develop a lot of things on my own, such as how to proceed with consultations, how to ask questions, and how to describe the medical records, and I visited various hospitals whenever I had the chance. What I found helpful in the descriptions in medical records was the medical records from the Department of Psychiatry. I was surprised to see that the conversations between doctors and patients were described, and I learned that the conversations between dietitians and patients are important for understanding patients and creating guidance policies. The 3 years I spent in the middle of Kabukicho, which is like a crucible where many different kinds of people live, allowed me to understand the difficulty and fascination of understanding human beings.

1.4 Training at Akasaka Szechuan Restaurant and Helping at the National Institute of Nutrition

1.4.1 Training in Chinese Cuisine

One day, Dr. Arai made a proposal. He wanted me to train as a cook at the Akasaka Szechuan Restaurant (赤坂四川飯店), which was run by Mr. Chen Jianmin (陳建民), who had introduced Mapo Tofu (麻婆豆腐) to Japan and appeared on NHK's "Today's Cooking". I was asked to become his apprentice. The restaurant was a training ground for chefs working at Szechuan restaurants throughout Japan, and talented chefs gathered from all over the country. For six months without pay, I was allowed to experience the cycle of washing dishes, pots and pans, and cutting boards, food preparation, seasoning, cooking and serving. Since work at the restaurant continued until around 10:00 p.m., it was late at night before I was taught by my seniors, and I spent many days sleeping at the restaurant. I became good friends with the people I worked with, and they accepted an oddball like me and taught me until midnight. However, in fact, the only thing that Mr. Chen taught me directly was how to peel bamboo shoots. When I was peeling bamboo shoots in a corner of the corridor, he told me directly, "You peel bamboo shoots like this". A fellow cook who worked with me commented , "You should become a chef because you have good cooking skills".

Later, when I talked to Dr. Arai. about it, he lumped it all together and said, "Becoming a chef is not your goal." I gave up the path of becoming a chef, but I remained close to my friends from those days.

I knew that I had to study food and cooking because people do not eat the nutrients, but eat what the ingredients are cooked with. The reason why I chose to study Chinese food was because Chinese food is based on the principle of "Yin Yang and the Five Elements (陰陽五行説)" in the selection of foods, cooking methods, and the relationship with the human body. I thought that there might be some similarity between Chinese cuisine and nutritional science because nutritional science is also classified into five kinds of nutrients. I studied Chinese medicine and medicinal herbs for a while, but there was a big gap between nutritional science, which was developed based on the elemental reduction theory of the components of food and the human body, and Oriental medicine, which classified foods based on the observation and experience of natural phenomena.

1.4.2 Helping Out at the National Institute of Nutrition

Another advantage of studying at the Shinjuku Clinic was that the National Institute of Nutrition (now the National Institute of Biomedical Research and Nutrition) was nearby and I had access to it. At that time, I was assisting the research of Dr. Shinjiro Suzuki (鈴木慎二郎) director of the institute, who was studying the relationship between nutrition and exercise for the first time. At that time, there was little basic research on obesity, and the relationship between diet and exercise was an important issue. I was responsible for the nutritional management, menu preparation, and cooking for four subjects. I prepared the meals. This experience gave me a firsthand understanding of the difficulties of human intervention research on nutrition. I became a frequent visitor to the National Institute of Nutrition, and was able to get to know many professors at the institute, which was very helpful in broadening my personal relationships.

1.5 I'm Glad I Learned About Nutrition

In 1975, I was approached by St. Marianna University Hospital. Shigeyoshi Saishoji, deputy director of the hospital, asked me if I would work with him to reform the hospital nutrition department. I worked hard to reform the hospital food service and establish a nutrition counseling room, and provided nutrition counseling to about 50,000 people until 2003, when the hospital was transferred to Kanagawa Prefectural University of Health and Welfare.

Ten years after I left the clinical field and moved to the educational field, I had an unexpected experience in the summer of that year. When I gave a lecture at the Yokosuka Labor Hall, a woman who wanted to meet me came to my dressing room after the lecture. She told me this story.

"I have finally gotten to meet RD. Nakamura. Actually, my husband was taken care of at the nutritional consultation room of St. Marianna University Hospital 25 years ago. After his stomach cancer surgery, he thought he needed to nourish himself in order to recover his strength quickly, so he tried to eat, but he could not eat as he wanted. His sense of taste had changed, he had no appetite, and since his stomach had been removed, he couldn't eat much at a time. At that time, RD. Nakamura helped us."

Her husband had passed away 2 years before, but he always looked forward to going to his nutritional consultations.

"At last, I have met you. Before he died, my husband left a request in his will that I must go to him to pay my respects. Now I can go to my husband with peace of mind".

It was a moment when I felt that the youthful dream of a young man who 50 years ago in a dimly animal room at my alma mater had realized the importance of nutrition and felt that he had to teach this to many people had finally been rewarded. I thought from the bottom of my heart that I was truly happy that I had chosen "nutrition" as my life's goal.

Chapter 2
The Birth of Nutrition and the Systematization of Learning

Abstract Nutrition is the process and state by which a living organism takes in food from outside the body, uses it, grows and develops, maintains life, and leads a healthy life, and the substances it takes in are called nutrients. The nutritional status of the human body can be roughly divided into deficiency and latent deficiency, and into excess and latent excess.

In 1953, the double helix structure of DNA was elucidated by Watson and Crick, and biochemical research focused on the genome and neglected nutrition which is an external factor. However, a person's individuality and diseases are not only determined merely by DNA but also related to the mechanism called the epigenome. Nutrition and biochemistry in the study of the mechanisms within cells are once again becoming more closely linked.

We have the eating habits of an "unbelievable omnivore". Each food provides some of the nutrients for human beings, but the composition of the food is not for the maintenance of human life and health. We began to seek wisdom about what to choose from such a wide variety of foods, how much to eat them. This is the reason why we need nutrition.

Keywords Nutrition · Nutrient · Nutritional status · Latent deficiency · Composition of the human body · Biochemistry · DNA · Epigenome · Omnivore · Homo sapiens

2.1 "Nourishing" Is Wrong

Emeritus Professor Norimasa Hosoya of the University of Tokyo defines "nutrition as the process by which a living organism takes in food from outside the body, uses it, grows and develops, maintains life, and leads a healthy life, and the substances it takes in are called nutrients". In other words, nutrition is the activity in which our body takes in food and processes it.

However, it is generally used in a confusing way without understanding the difference between "nutrition" and "nutrients". For example, it is often said that "spinach has nutrition", but this expression is not correct. Because spinach contains a lot of vitamins and minerals, but not nutrition.

T. Nakamura, *Japan Nutrition*, https://doi.org/10.1007/978-981-16-6316-1_2

2.1.1 Nutritional Value Depends on the Person

To be precise, spinach contains a lot of vitamins and minerals, so it is a nutritious food for persons who tend to be deficient in these nutrients. It is important to remember that the nutritional value of spinach depends on the type and amount of nutrients it contains, as well as the nutritional status of the person consuming it. Spinach is a valuable food for persons with a low intake of vitamins and minerals, but not for those who consume enough of these. If a person is thin and has a small appetite, rice and fat are better sources of energy than spinach, so these foods are more nutritious. It cannot be said that spinach is a valuable food for all people.

2.1.2 What Are Healthy Foods?

Before and after the war, many people were undernourished due to food shortages. At that time, foods that contained energy and various nutrients and were easy to digest and absorb were regarded as excellent foods with high nutritional value. Yam cakes, mushrooms, bamboo shoots, etc. were not considered to be valuable foods because their nutritional value was low due to their low content of various nutrients and the carbohydrates which they contained were dietary fiber without digestive enzymes. However, as obesity and metabolic syndrome became common and excessive energy intake became a problem, foods with low energy content and weak digestion and absorption were evaluated as valuable foods.

This is also true for individuals. I am often asked, "Is food X good for health?" To such a question, I answer, "It is good for those for whom it is effective." This is not a joke or an insincere answer, but my true meaning. If the ingredients of a certain food are effective in improving a person's current state of health and nutrition, it can be said to be said to be a healthy food for that person.

Improving health and nutritional status refers to the intake of energy and nutrients, etc., in order to maintain and improve health and keep people away from nutritional risks and disease. For example, if a person is obese and concerned about blood glucose and triglycerides, it makes sense to choose low-energy, low-fat, and low-sugar foods, but if a person is thin and prone to low blood glucose, these foods are not only meaningless but even unhealthy.

Choosing the right foods that form a healthy diet is not a matter of looking for "health foods" or "immortality foods" that work for everyone, but rather whether or not they are effective in improving one's health and nutritional status.

2.2 Nutritional Status of Humans

In order to select appropriate foods and supplements, the first step is to determine and assess the nutritional status of the user. Therefore, let us consider "what is nutritional status".

The components of the human body are constantly decomposed and replaced by newly synthesized ones as they become older. Moreover, some of them are decomposed and some are reused, but they are finally excreted through urine and skin. However, if the amount of nutrient intake is low for some reason, or if the amount of a nutrient required increases due to exercise or disease, and the level of degradation exceeds the level of synthesis, a nutrient deficiency will occur. If the deficiency is significant and prolonged, metabolic homeostasis cannot be maintained, leading to nutritional deficiency and disease. These changes start with biochemical changes in the cells, followed by physiological changes over time, and finally morphological changes in the tissues and organs. Such a comprehensive evaluation of changes in nutritional status is called nutritional assessment.

2.2.1 Nutritional Deficiencies and Excesses

The nutritional status of the human body can be roughly divided into deficiency and excess, with the former divided into deficiency and latent deficiency, and the latter into excess and latent excess.

Nutritional deficiency is a condition in which the deficiency of nutrients is prolonged and physical and mental abnormalities appear, such as beriberi, night blindness, scurvy, and rickets. Various kinds of nutritional supplements are required for treatment. Latent deficiency is a condition in which the body is not deficient in nutrients, but the nutritional intake is insufficient, and the body's storage of and metabolic capacity for nutrients are reduced, resulting in the appearance of various complaints. Since the body has a natural healing power, the deficiency can be improved by improving the daily diet and nutritional supplements.

Excessive nutrient intake is a state of addiction in which a large amount of a specific food or supplement is consumed, resulting in long-term excessive nutrient intake and the appearance of physical and mental abnormalities. In addition, it can result in non-infectious chronic diseases such as obesity, diabetes mellitus, dyslipidemia, hypertension, hyperuricemia, and arteriosclerosis, or so-called life-style-related diseases, due to the involvement of genetic predisposition to the excess nutrients. A potential excess state is a state in which various clinical laboratory values are not abnormal enough to be diagnosed as a disease, but the nutrient intake is excessive, the amount of body fat increases due to obesity, the metabolism of energy and nutrients is altered, and noncommunicable diseases: NCDs are easily induced. Metabolic syndrome is a condition in which the body fat, blood glucose, blood lipid, and blood pressure are above the standard values, but not high enough to be diagnosed as obesity, diabetes, dyslipidemia, or hypertension.

2.3   The Birth of Nutrition

From ancient times to the present, the relationship between diet and health or disease has been discussed many times. Various dietary methods have been proposed as methods to preserve health, maintain a regimen, and even prevent and treat diseases. Many of these traditional diets are based on the laws of human experience and observation of nature. For this reason, there was no concept that as ingested food is digested, certain components are absorbed, and that these components are related to the activities of life. Nutritional science, however, has developed physiology and biochemistry, discovered the components of life and growth in food, and clarified the universal relationship between food and life. I believe that nutrition is the study that has scientifically clarified the relationship between food and the activities of life. Therefore, it can be said that nutrition is the science that forms the basis of life science. Then, who gave birth to the concept of "nutrition"? In other words, who first came up with the idea of "nutrition"? To find the answer to this question, we first need to briefly unravel the history of science.

2.3.1   Nutrition and Life Sciences

At first, it was during the Greek era that human beings began to think deeply about the phenomena of the world. At that time, the so-called intellectuals discussed the philosophical issue of "What is a human being?" They were discussing a philosophical question. However, since human beings have many different aspects, no matter how many times they debated, they ended up arguing for the sake of arguing, and this did not lead to the progress of human beings.

In the early seventeenth century, an era of rapid academic progress arrived in Europe. René Descartes (1596–1650), a Frenchman who grew up in that era, was regarded as the founder of modern, rationalist philosophy. He separated the human body from the mind, making the human body the object of natural science, which pursues objectivity, and the mind the object of psychology, literature and art. He proposed the so-called human machine theory, which can be said to have laid the foundation for life science, which interprets life in terms of material change. In the life sciences, universality was emphasized, and issues of the mind and emotions were eliminated, leading to progress in anatomy, physiology, biochemistry, and molecular biology. Nutrition was developed as a part of these life sciences.

I believe that it was the discovery of combustion and energy metabolism that provided the starting point for nutrition to become an independent academic system within the life sciences. Human beings have long had an unusual interest in the phenomenon of burning. When something burns, it produces light that brightens up the world and heat that warms up our bodies, thus enriching our lives. Moreover, all animals fear fire and keep away from it, and only humans developed civilization by making and using fire.

2.3.2   The Founder of Nutrition

Until around the seventeenth century, it was believed that light and heat were produced by the release of phlogiston, the element that burns when an object burns. Lavoisier, a French scientist in the late eighteenth century, believed that combustion was a phenomenon in which metals reacted with oxygen, and that the respiration of animals and the combustion of things were similar. He proved that living organisms consume oxygen and produce carbon dioxide gas when they breathe, and that the amount of carbon dioxide gas is proportional to the heat produced, and he found that energy metabolism increases with food intake and exertion. It can be said that he is the "founder of nutrition" who opened the door of nutrition by proving that man obtains the energy for life from food. However, some say that he only studied respiration as a part of physiology and did not consider nutrition as an independent science.

2.4   Systematization of Nutritional Science

2.4.1   Calorimetric Studies

In 1866, Carl Feucht of Germany built a large calorimeter to directly measure human energy expenditure, and Max Rubner, a student of his, reported in 1883 that energy metabolism was proportional to body surface area. In 1902, Rubner laid the foundation for calculating the amount of energy consumed by calculating the physiological burning of carbohydrates, fats, and proteins, and he is considered by some to be the "founder of nutrition." The calorific value of nutrients is 4 kcal for carbohydrates, 9 kcal for fats, and 4 kcal for proteins per gram. This number is called "the Atwater calorie factor" and is considered to be the most important coefficient in nutrition. Even now, the "Atwater Award" is the most prestigious award of the National Society of Nutrition. This coefficient means that licking a mere 1 g of sugar grabbed with three fingers generates 4 kcal of physiological energy in the body. 4 kcal of heat is the amount of energy that can raise the temperature of 4 l of water by 1 °C. It can be understood that the human body produces energy efficiently from nutrients and uses a large amount of energy to maintain life.

2.4.2   Carbohydrate and Fat Research

In the nineteenth century, the digestion of carbohydrates was elucidated and various digestive enzymes were discovered; in the early twentieth century, research on the metabolism of absorbed carbohydrates began, and in 1937, Hans Krebs (Germany) discovered the TCA cycle, in which carbohydrates are glycolyzed and oxidized to carbon dioxide and water to produce energy. The oxidation of lipids to produce energy was explained. Justus Liebig et al. (Germany) discovered that fats are synthesized from other nutrients, and that fats are not only a source of energy but also contain essential fatty acids that are involved in physiological functions such as growth, reproduction and formation of skin.

2.4.3   Protein Research

In the nineteenth century, full-scale research on proteins began, and it was found that the nutritional value of proteins was related to the amount of nitrogen contained in foods. In the twentieth century, it was confirmed that proteins are composed of amino acids, and it was clarified that the quality of proteins is determined by their amino acid composition. Subsequently, research developed into the classification of essential amino acids, which are not synthesized in the body, and non-essential amino acids, which are synthesized in the body, the requirements for amino acids, the balance of amino acids, and the physiological effects of various proteins and amino acids.

2.4.4   Vitamin Research

In the late nineteenth century, it was discovered that nutrients included not only the energy-producing nutrients of carbohydrates, fats, and proteins. There was also speculation about the existence of micronutrients. For example, in Japan, people who had the habit of eating too much white rice suffered from an intractable disease with neurological symptoms, and many soldiers died of this disease during the Sino-Japanese War and the Russo-Japanese War. Later, it was discovered that the disease was caused by a deficiency of vitamin B_1, which led to the discovery of the existence of micronutrients. The Army believed that the disease was an infectious disease, and enforced hygiene, but because this did not improve the diet, they were unable to reduce the number of patients, and the number of deaths from the disease was four times that of the actual war dead. The Navy, on the other hand, prevented beriberi from early on by switching from a diet centered on white rice to a Western diet centered on meat, as there were no cases of beriberi in the West.

In 1890, Eijkman (Netherlands) discovered that rice bran could be added to the feed of chickens showing symptoms of beriberi, and in 1911, Funk (Poland) succeeded in crystallizing the active ingredient from rice bran, which had the properties of an amine, and named it the amine of life, or "vitamin". In Japan, too, Dr. Umetaro Suzuki (鈴木梅太郎) succeeded in crystallizing the active ingredient from rice bran and clarified that beriberi was caused by a vitamin B_1 deficiency.

In Europe as well there were great nutritional accidents and discoveries. From the sixteenth and eighteenth centuries, some two million sailors died of a mysterious disease. In 1747, Captain James Cook (England) followed the advice of Dr. James Lind and prevented the disease by feeding his sailors citrus fruits, a local folk remedy at the time. The disease was scurvy, a vitamin C deficiency caused by a lack of fresh fruits and vegetables. In other words, at that time, Asia was suffering from a vitamin B_1 deficiency, while Europe was suffering from a vitamin C deficiency.

2.4.5   Mineral Research

In the eighteenth century, iron was found to be contained in blood, and bones were found to be composed of calcium and phosphorus. In the twentieth century, various mineral deficiency diseases were discovered, and the physiological effects of minerals and their content in foods, for example that goiter is caused by iodine deficiency were explained. Since then, many vitamins and minerals have been discovered.

When we look at the history of the development of nutritional science, we can see that first the nutrients that produce energy for life were discovered, and then the nutrients that make up the human body and carry out life activities were discovered, and the framework of the current five major nutrients was established. Finally, the characteristics of the foods containing these nutrients, a rational manner of intake, and the deficiency and excess of nutrients were discovered, and the relationship between food and drink and health and disease could be interpreted scientifically, and the academic systematization of nutrition was prepared.

2.5   Composition and Nutrition of the Human Body

Since nutrition is the scientific study of the relationship between the human body, food, and diet, we will first consider nutrition from the perspective of the human body.

2.5.1   Changes in Nutrients in the Human Body

The human body is composed of cells, tissues, and organs, and life is maintained by these interacting and working smoothly. 16.4% of the human body is protein, 15.3% is fat, 5.7% is mineral, less than 1% is sugar, and the rest is water. On the other hand, carbohydrates, lipids, and proteins constitute 57.7%, 26.3%, and 16.0% of the daily diet, respectively.

In other words, the human body is composed of nutrients as well as food, but the nutritional components of food are not directly utilized by the living body. The absorbed nutrients are mainly converted into suitable new nutrients in the liver, then stored, and finally circulated throughout the body to supply the needs of each organ and tissue. For example, when a person eats steak, which is the muscle of a cow, the protein ingested does not directly become the protein that makes up the muscles of a person. The protein in the steak is digested by digestive enzymes and absorbed in the form of amino acids, which are then combined with amino acids broken down from the proteins that make up the human body and are used to synthesize the protein needed by the person.

The blueprint used for this synthesis is the gene inherited from one's parents, and the protein suitable for the person is manufactured based on this information. If beef proteins were used as they are, the muscles of people who eat steaks would be the same as those of cows, but this is not the case. In other words, the proteins contained in the food we consume have species-specific characteristics, but if they are broken down to amino acids, the specificity of beef will disappear in the human body.

2.6   Nutrition and Biochemistry

Recent developments in molecular biology have revealed the metabolism and action of nutrients at the cellular and genetic levels. For example, a cell has a cytoplasm and a nucleus, and the cytoplasm includes mitochondria, lysosomes, an endoplasmic reticulum, and a Golgi apparatus. Mitochondria produce ATP, a component of energy, lysosomes produce proteins, the endoplasmic reticulum transports substances, and the Golgi apparatus encases proteins produced by ribosomes and transports them out of the cell. The synthesis, degradation, and metabolism of nutrients are now understood at the cellular level.

2.6.1   Conflict Between Nutrition and Biochemistry

Let us consider the relationship between cells and genes and also nutrition. Traditionally, nutrition and biochemistry were on a friendly honeymoon, and it was the biochemists who led the basic research in nutrition. In 1953, the double helix structure of DNA was elucidated by Watson and Crick, and biochemical research focused on the genome, which is an internal factor of the human body. After the elucidation of the double-helix structure of DNA by Watson and Crick, biochemical research became focused on the genome, and neglected nutrition, which is an external factor. The study of basic nutrition came to be treated as a minor part of biochemical research. At that time, nutritional deficiency caused by the food shortages due to World War II had been solved, signboards for the "Nutrition Laboratory" disappeared from the laboratories of universities in Japan, because the original research goal of nutrition became diluted. Nutrition also disappeared from medical education and training, nutrition researchers began to call themselves biochemists, and nutrition lost its luster.

However, by the end of the twentieth century it became clear that there is a mechanism that determines the function of genes regardless of the DNA base sequence. The sum total of such information is called the epigenome. In other words, it is now known that a person's individuality and diseases are not determined merely by the content of his or her DNA, but are also related to the mechanism called the epigenome. In the nucleus of a cell, the 46 chromosomes inherited from both parents are stored; the chromosomes consist of a double-helix structure of deoxyribonucleic acid (DNA) linked in chains. When the genetic information incorporated in DNA is read for physiological functions, it is called gene expression. In gene expression, the information in DNA is copied into messenger RNA, and amino acids are combined in the ribosome according to this information to synthesize the necessary proteins. In other words, the genetic information from the parents is transmitted from DNA to mRNA to protein, resulting in a body that resembles the parents, but is unique to the individual.

However, about 60 trillion cells in the whole body contain genes in almost the same way, but each of them develops into a component of a unique organ, and some cells become legs, and some cells become skin. This is because there is a device that turns on and off the action of genes in DNA, and it was found that this is controlled by various chemical modifications (methylation and acetylation) that bind to histone proteins that DNA is wrapped around.

Although the same genome is contained in all the cells that make up the human body, they can become various types of cells. In addition, identical twins have the same genome, but their appearance is similar but not identical, and if they grow up in different environments, they may have different body shapes, develop different diseases, and have different life spans. In other words, this reveals that human individuality and health conditions are determined not only by genetic information in the genome but also by differences in the epigenome. And when overeating, unbalanced nutrient intake, and stress are added, changes and abnormalities occur in this regulatory function, resulting in differences in protein production and function, the emergence of individual conditions, the inability to maintain health, and consequently, disease.

2.6.2 Linkage Between Nutrition and Biochemistry

Nutrition, the study of the external environment of the human body, and biochemistry, the study of the mechanisms within cells, are once again becoming more closely linked. When genes are compared among individuals, their nucleotide sequences may differ, and when these changes occur at a frequency of 1% or more of the population, they are called genetic polymorphisms. It has been found that if inappropriate dietary habits are followed by people with such genetic polymorphisms, they are more likely to develop diseases. Therefore, genetic diagnosis can reveal the constitution of an individual that makes him or her susceptible to developing a disease, which can be more reliably prevented if that person works to improve his or her eating habits from an early stage so that gene expression does not occur. For example, if there is a polymorphism in the gene for diabetes, the secretion function and sensitivity to insulin will be reduced, and the person will be prone to diabetes even if he or she is not overweight. Therefore, it is important to control the condition by losing weight, and if this is done, the onset of diabetes can be prevented more reliably. In other words, it is beginning to be understood that nutrients and diet play a role as the materials that make up the organism, and at the same time, affect the action of genes, which are the blueprint of the organism.

2.7 Advances in Nutrition and the Human Diet

Nutrition has been around for almost 200 years, since Lavoisier opened its doors. During this time, remarkable progress has been made. In medicine, central venous nutrition, which uses a tube to supply nutrients directly into the bloodstream, came into use, saving the lives of patients who had lost the function of their digestive organs, such as those with Crohn's disease. Humans now have a way to live without food, by administering nutrients directly into the body. Since infusions can contain only those nutrients that are known to man, what this method of nutritional supplementation alone means is that we have discovered almost all the nutrients that are the building blocks of life, meaning that we can live for decades.

If we draw this conclusion, we can live if we take all the necessary nutrients in pill form as a supplement every day. We will be free from troublesome shopping and cooking, and most of all, we will not need agriculture. People will be freed from food crises, starvation and malnutrition. When I first became interested in nutrition, I dreamed of a society in which people could wake up in the morning, throw a few supplements in their mouths, and not have to eat.

2.7.1 Development of Complete Nutritional Foods

In fact, there are people who have made this dream come true. They were the nutritionists who participated in the Apollo missions that landed man on the moon for the first time.

In 1970, three astronauts aboard Apollo 13 returned to Earth within two weeks of arriving on the Moon. What they ate during that time was "space food," a complete nutritional food that could keep them alive if they ate only that, and did not produce any stools. At the time, the American nutritionists who conducted the research were praised worldwide, and it was reported as a victory for American nutrition. I remember thinking with regret that I had been beaten to the punch.

However, further development of this space food was later frustrated. The reason for this was that the space pilots complained that if they were forced to eat something like toothpaste in cheesecloth at every meal, the food would become stressful and interfere with their work.

Whether human beings have discovered all the nutrients or not is not the issue, but there is a big difference between drinking or consuming all the nutrients from a tube and eating food prepared from nature. In addition to nutritional support, meals have the significance of providing the color, texture, aroma, and taste of the food, and are related to the production, distribution, processing, and cooking of the food, as well as the local culture, economy, and climate that make these processes possible. Since nutrition science is based on modern rationalism and uses elemental reductionism as its methodology, it tends to lead to the final conclusion of "rational, convenient, and inexpensive supplementation of nutrients," but we must not forget that diet is formed from a wide variety of factors and purposes.

The process of production, distribution, processing, and cooking is becoming more and more rationalized and simplified by the use of AI and robots, not only for space food or nutritional support for the sick and injured, but also for daily meals, and is moving farther and farther away from the "heartfelt meals prepared with time and effort" that humans originally enjoyed. This direction is inevitable in modern civilized society, which is based on modern rationalism, and its original sin may be found in Descartes' philosophy, which divided human beings into the human body and the mind, and became the foundation of modern science. However, it is precisely because we live in such a society that I believe we must develop "human nutrition," a field that explores the nature of nutrition and diet in a way that is closer to human beings.

2.8   Human Evolution and Nutrition

2.8.1   The Need for Nutrition Science

I have been able to understand the birth of the concept of nutrition and the development of nutritional science by exploring its history, but why, in the first place, humans need knowledge of nutrition science. Pandas eat only bamboo leaves and koalas eat only eucalyptus leaves, but their muscles are well developed and they live well. Even if we don't limit ourselves to such herbivores, there are no animals other than humans that eat with any consideration for nutrition. Even so, they grow normally and live their whole lives. Why can't humans live without thinking about this and that and eating?

I believe that the reason for this lies in humans having the eating habits of an "unbelievable omnivore". What we call food in our daily lives are animals and plants that exist in the natural world. However, these animals and plants do not originally exist to be used as food for human beings. Each food provides some of the nutrients necessary for human beings, but the composition of the food is for the animals and plants to survive, not for the maintenance of human health.

For example, pork contains a lot of protein and vitamin B_1, but it is also a food with a lot of saturated fatty acids, so it is an excellent food for supplying protein and vitamin B_1, but eating too much of it can lead to obesity and dyslipidemia. In fact, there is no "complete nutritional food" in the natural world that contains all the nutrients necessary for humans. Humans have never encountered a "so-called health food" like bamboo leaves for pandas, which could keep us alive if we eat only that. Therefore, humans chose to be omnivorous, which means that we combine a variety of incomplete foods and consume the necessary nutrients in a well-balanced manner throughout our diet. Rather than a conscious choice, it is more reasonable to think that only omnivores survived and evolved into humans. In other words, in the process of evolution, human beings encountered severe food shortages many times and had to eat everything, and each time that they increased the variety of foods they ate, they expanded their omnivory and acquired the adaptability to live anywhere on earth.

As human beings developed their culture and civilization and wished to be healthier and live longer, they began to seek wisdom about what to choose from such a wide variety of foods, how much to eat, and how to cook and consume them. It is my understanding that nutritional science is the scientific elucidation of this wisdom.

2.8.2   Acquisition of Omnivorousness

So what led humans to become omnivorous?

I looked into the process of human evolution. The common ancestor of the present great apes existed about 5 million years ago, from which the orangutans and gorillas were separated, and then about 70 million years ago, the chimpanzees and hominids were separated. We are Homo sapiens was the only one that survived. The other hominids became extinct due to diseases, environmental changes, and predation. Why did we alone survive? There are many theories.

It has long been theorized that the human race evolved from hanging from trees to moving from the jungles of Africa out onto the grasslands, evolving bipedalism, developing large brains, and expanding their diet. However, several species with large brains appeared among the extinct hominids. For example, Homo sapiens' biggest rival, the extinct Neanderthal man, had a brain capacity of about 1550 cc, while Homo sapiens had a smaller brain capacity of about 1450 cc. Incidentally, the brain of modern humans is even smaller at about 1350 cc. Moreover, the brain consumes as much energy as muscle, and a large brain requires more energy and increases the basal metabolic rate. There is a study that calculated the basal metabolic rate, and according to the study, the basal metabolic rate of Neanderthal man was 1, 2 times higher than that of Homo sapiens. Neanderthals, who were large and robust, needed a lot of energy and food to survive, and were not suited to an environment where food was scarce.

2.8.3 Characteristics and Evolution of Homo sapiens

Homo sapiens was not physically strong compared to other human species; rather, it was stocky and lacked arm strength. However, Homo sapiens formed a monogamous family early on, and males distributed food to females and children. As a result, it became necessary for them to harvest and carry food not only for themselves but also for their families, and they evolved to walk on two legs in a straight line and use their hands to carry large amounts of food. Molecular paleontologist Isao Sarashina refers to this as the "food transport hypothesis". Because upright bipedal walking was a rational way of moving with little physical strain, humans were able to obtain food from distant places and increase its variety. They actively left the jungle and began to eat leftovers from carnivores. These animal foods were easy to digest and rich in high quality proteins, fats, vitamins and minerals, which made it easier for humans to digest and develop their brain functions as well as their brain size. Herbivores eat cellulose from plants, and microorganisms in the digestive tract produce various active ingredients through fermentation, which they absorb and use as nutrients, but with this method, they have to eat a large amount of plants and constantly chew, swallow, and digest.

By developing brain functions, Homo sapiens used fire, advanced hunting techniques, and agriculture, evolving to be increasingly omnivorous and expanding their living area to every corner of the earth. For example, our greatest rivals, the Neanderthals who evolved in Europe, were robust, smart, and active, but they ate only a limited range of foods. In human terms, they were "picky eaters". As the Earth entered a cold period and the flora and fauna dwindled from the land masses of Europe, the Neanderthals lost their food supply and became extinct, with the Gibraltar Peninsula being the last place they lived. On the other hand, Homo sapiens, which evolved in Africa, survived by expanding its omnivorous diet to include fish, shellfish, and seaweed when food became scarce on land during the cold season.

In the process of evolving the function of upright bipedal walking, humans expanded their omnivorous nature and urvived the harsh changes in the global environment. It would be better to say that humans did not choose to be omnivorous, but that only omnivores survived. However, this omnivory has left us with an important problem. This is the need to make proper food choices from a wide variety of foods, and nutrition was the answer to this problem. In world history, there have been several cases of groups of people who, for some reason, lacked certain foods or ate too much of certain foods, and when they could not maintain a proper omnivorous diet, their health was damaged, they died in large numbers, and they became almost extinct.

"What is a proper omnivore diet?"

To put it simply, it is "a diet in which all nutrients are consumed without excess or deficiency." Specifically, it is a diet in which carbohydrates are consumed from grains such as rice, bread, and noodles; proteins and fats are consumed from meat, seafood, eggs, and soy products; and vitamins and minerals are consumed from milk and dairy products, as well as from vegetables and fruits.

As mentioned above, our ancestor, Homo sapiens, evolved into modern humans by choosing to be omnivorous, eating all kinds of plants and animals in order to survive in a harsh environment. The industrial revolution, which took place in Europe in the eighteenth century, enabled a cultured and affluent diet and created a society in which the population grew remarkably and people could live longer. This was supported by the "science of nutrition" that was born in that era.

At present, "nutrition" has become a subject of systematic research and education and has created an independent academic system of "sciences of nutrition". It consists of "basic nutrition" that deals with the basic matters of nutrition, "applied nutrition" that applies and practices nutrition in each stage of life, "food nutrition" that focuses on food, "clinical nutrition" that focuses on the individual human being, and nutrition for public health that focuses on groups or society. In addition, there are "nutrition education theory" and "food service management theory" as methodologies to put this knowledge into practice.

Bibliography

Dries R, translated by Ogawa T (2016) Food and humanity. Nihon Keizai Shimbun
Grotzer W/Shigeki Kitagami Translation (2008) History of nutrition. Kodansha Scientific
Harada N (2008) What's eating? Chikuma Primer Shinsho
Harari YN, translated by Shibata H (2016) The prehistory of sapiens, Vol. 1 and 2. Kawade Shobo Shinsha
Hatanaka S (2017) Charisma food. Shunjyusha

Hosoya N (2000) Sankai Human Nutrition. Cooking and Nutrition Education Corporation
Japan Dietetic Association, Nutrition Guidance Research Institute, Health and Nutrition Informa-
    tion Study Group (ed) (1998) Postwar Showa era nutrition trends: looking back on 40 years of
    the national nutrition survey. Dai-ichi Shuppan
Koike G. Nutritional science no nai tachi to purposu. Syst Nurs Course:2–6
Koyamano A (2013) An introduction to world history for Japanese. Shincho Shinsho
Muroteinas L, translated by Mizutani J (2016) Chapter 8, What is matter made of, "Four million
    years of humanity and science to understand this world". Kawade Shobo, pp 194–237
Oiso T (1977) What about population, food and nutrition. Dai-ichi Shuppan
Sarashina I (2018) Human history of extinction: why 'we' survived. NHK Shuppan Shinsho
Sato Y (2016) Human history of food. Chukoh Shinsho
Shopman P, translated by Kawai N (2015) Humans and dogs exterminated Neanderthals. Hara
    Shobo
Wurter C, Nagano T, Akamatsu M (2014) Seven million years of human evolution. A story.
    Seidosha

Chapter 3
History of Nutritional Improvement in Japan

Abstract Japan has solved the problems of nutritional deficiency due to shortage of food before and after the war II, and overnutrition associated with westernization after high economic growth. Japan has solved this double burden of malnutrition and has become to be a healthy longevity country. To find out why Japan has succeeded in improving nutrition, it is necessary to study the history of Japanese nutrition.

Nutritional science was introduced from the West after the Meiji Restoration. The Meiji government proclaimed, "wealthy country, strong military" and "promotion of industry and commerce" and focused on nutrition in order to improve the physical condition of the people. The modernization of the Japanese diet did not result in the exclusion of Western food, but rather in the formation of a new food culture through the fusion of Western food with Japanese food.

Dr. T. Saeki felt that it would be difficult to improve the nutritional status of the people only by providing information by nutrition researchers. He established the "Nutrition School" to train nutrition professionals. Dietitians were born in the "Dietitian Act" in 1948, they were assigned to group meal facilities to provide meals with excellent nutritional balance while utilizing the limited amount of food available, and to ensure nutrition education. Using facilities were schools, industrial concerns, hospitals, welfare facilities.

The "Kitchen Car" in which the rear part was converted into a kitchen for cooking demonstration, and dietitians boarded and provided nutrition education to every corner of Japan.

The children's physique has improved remarkably as a result of the school lunch program. Top: immediately after the start of the school lunch, middle: 4 months later, bottom: 2 years later (numbers indicate the same child.)

Keywords Meiji restoration · The modernization · Western food · Beriberi · Basic Law on Food Education · Sino-Japanese War · Russo-Japanese Wars · National Institute of Nutrition · Nutrition School · Dietitian Regulations · Nutrition Improvement Act · Nutritional guidance · Kitchen Car · Food truck · Food policy · Nutrition policy · National Nutrition Survey · Japan Dietetic Association · Dietitians Act · Dietitians · Registered dietitian · National Training Seminar on Pathological Nutrition · Non-communicable chronic diseases: NCDs

© The Author(s) 2022
T. Nakamura, *Japan Nutrition*, https://doi.org/10.1007/978-981-16-6316-1_3

3.1 Introduction to Nutrition

We decided to unravel the history of nutrition in Japan. It is important to preserve for future generations what we have done, and it is also helpful in creating a vision for the future. There are several descriptions of the history of nutrition and diet. Most of them are deterministic, and few of them go into the social background or the thoughts and efforts of the people and groups involved. However, history is called history. In other words, it is his (he's) story (Story), and since it is greatly influenced by the talents of outstanding people, we have decided to approach the personalities who moved history. This is because I thought that this would be helpful for young people who will be responsible for nutrition in the future. If Dr. Tadasu Saiki, Dr. Kanehiro Takaki, Mr. Kiku Morikawa, or Dr. Norimasa Hosoya had not appeared in their times, the history of nutrition in Japan would have been completely different.

3.1.1 The Concept of Chinese Medicine

It was after the Meiji Restoration that nutritional science was introduced to Japan. Before the Meiji Restoration, Chinese medicine was the foundation of medical care in Japan. Chinese medicine is classified into drug therapy, acupuncture and moxibustion therapy, and dietary regimen. However, the purpose of dietary regimen was not to consume energy and nutrients properly as in current dietetic treatment, but to systematize food selection and cooking methods based on the theory of yin-yang and five elements. Moreover, the fundamental feature of dietary regimen was that "medicine and food have the same source", which means that medicine and food were interpreted as one and the same thing, and the efficacy of food as a whole is organized according to its dietary function, taste, and nature. Food has five major tastes: sour, bitter, sweet, spicy, and salty, and each taste is considered to have a specific effect (Table 3.1). On the other hand, in nutrition, nutrients are divided into five groups according to the characteristics of their constituents and their roles in the body (Table 3.2). It is interesting to note that the characteristics of both Chinese Medicine and nutrition science were classified into five groups.

Diet regimen is based on categorizing the characteristics of foods based on long years of dietary experience, and selecting foods according to the user's signs (constitution). For example, the body's constitution is largely divided into two types: the deficiency-cold type and the actual-heat type, and the foods to be taken and the foods to be avoided are thereby determined. For example, in the case of a

Table 3.1 Foods with five flavors and their effects (Chinese medicine)

1	Acidity	Lemon, Japanese apricot, tomato, plum, yogurt, etc.
		It has astringent and anti-inflammatory effects, and is effective for night sweats, diarrhea and polyuria.
2	Bitterness	Celery, green pepper, coffee, tea, seaweed, etc.
		It has the effect of drying and firming water retention and is effective for fever.
3	Sweetness	Cereals, potatoes, eggs, milk, meat, fish, fruits, vegetables, etc.
		It has the effect of nourishing the weakness of the human body and relieving pain, and is effective in nourishing deficiency of blood.
4	Spicy	Leek, radish, garlic, chives, ginger, pepper, sansho (Japanese pepper), etc.
		It warms and dilates the body and improves the circulation of qi and blood.
5	Salty	Salt, barley, soy sauce, miso, pickles, salted fish and meat, etc.
		It has a soothing and moistening effect and is effective for lumps under the skin, swelling of the Lymph is a fluid and constipation.

Table 3.2 Five nutrients, food groups and their actions in nutritional science

Energy-producing nutrients	
1	Proteins: Meat, seafood, eggs, soy products
	A source of amino acids, which are the building blocks of the body
2	Lipids: Fats and oils
	A source of fatty acids, an efficient source of energy and a functional ingredient
3	Carbohydrates: Grains, potatoes
	Carbohydrate that is digested and absorbed as an energy source and has physiological functions and food function as fermentable energy
Micronutrients	
4	Vitamins: Milk and dairy products, vegetables, fruits
	Organic compounds that are not synthesized or insufficiently synthesized by the body to regulate metabolism
5	Minerals: vegetables, fruits, milk and dairy products
	Inorganic compounds that act to regulate metabolism and are components of the body.

cold-related deficiency – the body wouldn't be deficient in cold, the foods that possess the heat-supplementing property, such as those that are warm and those that are sweet and spicy in taste, should be taken, while in the case of the actual heat, the foods that that have the purgative cool property, such as those that are cold and those that are sour and bitter in taste, should be taken in relatively large quantities. In the Edo period and the early Meiji period, this type of dietary cure was used as a dietary therapy and dietary indicator by the Japanese. For example, "eat eel on Saturdays" and "take porridge and pickled plums when you are sick" are some examples of such dietary wisdom.

3.1.2 Anglo-American Nutrition and German Medicine

Modern nutritional science was introduced from the West after the Meiji Restoration. Nutritional science does not merely classify foods as in Chinese medicine, but analyzes the ingredients in foods that are effective for the organism and selects foods according to their contents. In 1859, Dr. Hepburn (James Hepburn) came to Japan from England and opened a clinic in the port of Yokohama, which had been opened to foreigners. He was known as the founder of the "Anglo-American Yokohama School" because he introduced modern medicine to Japan. His students included Mr. Yukichi Fukuzawa and Dr. Kanehiro Takagi, who later contributed to the development of clinical nutrition in Japan. On the other hand, in 1887 (Meiji 20), by a resolution of the Diet the Japanese government introduced German medicine in order to modernize medicine. At the time, German medicine was based on experimental medicine, which put it at odds with the Anglo-American Yokohama School, which was based on positivism and insisted on the importance of clinical medicine. In fact, the subsequent controversy between Dr. Rintaro Mori (Army), in the mainstream of German medicine, and Dr. Kanehiro Takagi (Navy), who had studied British medicine, seems to have been based on this fact. In other words, practical nutrition in medicine was influenced by Britain and the United States, while basic nutrition as an academic discipline was influenced by German medicine. Dr. Feucht, a German physician, gave a lecture on the concept of nutrition to the students of the Faculty of Medicine at the University of Tokyo, saying that food should not be eaten according to taste, but according to the nutritional components it contains.

3.1.3 A Fusion of Traditional Japanese Food and Western Food

The Meiji government, with its goal of establishing a modern nation, proclaimed "wealthy country, strong military" and "promotion of industry and commerce" as policies and focused on nutrition in order to improve the physical condition of the people. The introduction of nutritional science and nutritious Western food was a way to improve the health of the people. The westernization of dietary habits was essential as a national policy, and nutrition was actively utilized as the logical basis for this policy. A symbol of the modernization of the diet was the "encouragement of meat-eating". Japanese people's preference for abstinence from meat-eating dates back to the time when their food culture was originally centered on rice cultivation, and when Emperor Temmu ruled 673–686, under the influence of Buddhism, issued a decree banning meat-eating, which lasted until the Edo period. In the latter half of the Edo period, however, trade with foreign countries increased, and places influenced by Western cuisine gradually emerged in urban areas and open-port regions, with a beef stew shop opening in Yokohama in 1862 and a butcher shop in Takanawa, Edo in 1867. Later, Western cuisine was introduced through cookbooks, newspapers, and magazines, a Japanese-Western eclectic cuisine, which combined Japanese and Western food, was born and gradually spread.

The modernization of the Japanese diet did not result in the exclusion of Western food, but rather in the formation of a new food culture through the fusion of Western food with Japanese food. The Westernization of the diet, which was recommended by the government as a national policy, was a result of the traditional culture of rice cultivation, the climate of Japan, and the ingenuity of the common people, and it had the advantage of improving nutrition by providing energy, protein, fat, vitamins, and minerals, which had been in short supply.

3.2   The Beriberi Controversy and the Basic Law on Food Education

3.2.1   Longing for White Rice and Beriberi

For the Japanese people, a full meal of white rice was a dream come true. Rice was a crop that had to be paid as annual tribute and was not easily available to the common people. Even among the wealthy farmers of the countryside, rice was used only mixed with minor grains, and the common people ate sweet potatoes, barley, and Japanese millet as their staple foods. This means that Japanese people originally ate a diet based on minor grains in general. In 1873 (Meiji 6), when the Meiji government changed the taxation system to payment in money, rice was left in the hands of the farmers. At the same time, improved rice cultivation increased production and the development of the just 'silk' industry provided a cash income with which to purchase rice. The consumption of rice increased, and a new staple food-oriented eating pattern was formed. Furthermore, with the progress of threshing technology, the dream of white rice became a common food.

Ironically, this widespread dependence on white rice led to an epidemic of beriberi, which became a problem, especially in the military. White rice was used in advertising campaigns to recruit men to join the military, with the promise that recruits could eat as much white rice as they wanted. However, Dr. Kanehiro Takagi, who had studied at the Anglo-American Yokohama School and became the navy's medical superintendent, claimed that this diet was the cause of beriberi. He had studied epidemiology in England and believed that the reason why Europeans did not suffer from beriberi was the difference in diet, and he changed the diet on warships from Japanese to Western.

In contrast, Dr. Rintaro Mori, an army doctor who had studied under Dr. Robert Koch, a bacteriologist who had followed the mainstream of German medicine introduced by the Meiji government, directly opposed Takagi's theory, arguing that the cause of beriberi was a microbial infection.

The beriberi controversy between the Army and Navy developed into a beriberi war between the two branches of the military.

3.2.2   The Army and Navy Legume Controversy

In December 1882 (Meiji 15), the 371 crew members of the naval training ship "Ryujo" sailed from Japan to New Zealand, Chile and Peru, returning home after a voyage of 272 days. Among the crew there were 160 cases of beriberi and 25 deaths from it. During the voyage, the sailors continued to eat a Japanese diet consisting mainly of white rice. The navy was so shocked that it rushed to introduce Western-style food, and Takagi conducted a large-scale clinical experiment to prove his point: for 187 days beginning in February 1884 (Meiji 17), he had the warship "Tsukuba" sail along the same route as the "Ryujo", the ship that had suffered the attacks. During this time, the ship's crew was fed a completely Western diet of barley rice, meat, condensed milk and biscuits. As a result, of the 333 crew members on board, the number of people who developed beriberi was reduced to six and the number of people who died of beriberi to zero. The Navy made similar changes to the diet on other warships and at shore facilities. In 2004 (Heisei 16), Dr. Hidenori Kido and his colleagues analyzed the data from 16 warships and 8 shore facilities using a new statistical method. As a result of the change, the incidence of beriberi in all facilities decreased from 27% to 14%, and the odds ratio was significantly reduced to 0.38 by meta-analysis.

In 1894 (Meiji 27), during the Sino-Japanese War, the government argued for a change to a Western diet of barley rice, but this was not implemented due to opposition from senior army officials. The following year, Dr. Rintaro Mori wrote a book entitled "The Theory of Food for Japanese Soldiers", in which he criticized Takagi's improvement of diet for sailors. Since no conclusion was reached to this conflict, it was eventually resolved in the Sino-Japanese and Russo-Japanese Wars. During the war, the number of people who developed beriberi in the Army was about 1200 times greater than in the Navy, and the number of people who died of beriberi was about 4000 times greater. Incidentally, the number of deaths in the Army from beriberi was four times the number of those killed in action, meaning that many soldiers in the Army died from malnutrition rather than from guns (Table 3.3).

Table 3.3 Occurrence of beriberi during the Sino-Japanese and Russo-Japanese Wars

	Army	Navy
Numbers of Troop	240,616 people	3096 people
Deaths in action	1132 people	337 people
Numbers of Beriberi patients	41,431 people	34 people
Deaths from beriberi	4064 people	1 people

In 1913 (Taisho 2), the Army changed its ration to seven parts white rice and three parts wheat, and after that, the number of patients with beriberi decreased dramatically. Thus in 1924 (Taisho 13), the Provisional Investigation Committee concluded that "beriberi was caused mainly by vitamin deficiency". During this period, Dr. Umetaro Suzuki isolated and crystallized an effective ingredient for the prevention of beriberi from rice bran, and this played an important role in the beginning of vitamin research. In other words, we have entered an era in which nutrients are not only energy sources, but also include micro-nutrients, vitamins and minerals, that regulate our physical condition.

3.2.3   Promulgation of the Nutrition Education Basic Act

By the way, in 2006 (Heisei 17), the Japanese government promulgated the Nutrition Education Basic Act as the world's first national basic law. It states that dietary education, which is the foundation of intellectual and physical education, is necessary for raising healthy children (Table 3.4). Every month for a year, a review meeting was held at the Prime Minister's Office, and I participated as one of the members (Table 3.5). The members of the review committee were outstanding, and

Table 3.4 Nutrition Education Basic Act

For the development of Japan in the twenty-first century, it is important to cultivate sound minds and bodies in children so that they can spread their wings toward the future and the international community, and to ensure the physical and mental health of all citizens so that they can live vibrant lives throughout their lifespan. In order for children to develop a rich sense of humanity and acquire the ability to live, food is of the utmost importance. In addition, it is necessary to promote dietary education so that children can acquire knowledge about food and the ability to choose food through varied experiences, and so that they can practice sound dietary habits. While dietary education is necessary for all generations of citizens, dietary education for children has a significant impact on their physical and mental growth and the formation of their personalities, and serves as the basis for cultivating a sound mind and body and nurturing a rich humanity throughout their lives.

Table 3.5 Study Committee Members for the Nutrition Education Basic Act

Chairman
Junichiro Koizumi Prime Minister

Committee members (25)	
Hiroyuki Hosoda	Chief Cabinet Secretary
Yasufumi Tanahashi	Minister of State for Special Missions (Food safety and nutrition education)
Taro Aso	Minister of Internal Affairs and Communications
Chieko Noono	Minister of Justice, Minister of State for Special Missions Youth development and measures for declining birth rates
Nobutaka Machimura	Minister for Foreign Affairs
Sadakazu Tanigaki	Minister of Finance
Nariaki Nakayama	Ministry of Education, Culture, Sports, Science and Technology
Hidehisa Otsuji	Minister of Health, Labour and Welfare
Mineichi Iwanaga	Minister of Agriculture, Forestry and Fisheries
Shoichi Nakagawa	Minister of Economy, Trade and Industry
Kazuo Kitagawa	Minister of Land, Infrastructure and Transport
Yuriko Koike	Minister of the Environment
Yoshitaka Murata	Chairman, National Public Safety Commission

Committee members (25)	
Shoko Ichiba	Vice President, Japanese Association for Dietetic Research and Education
Iccho Itou	Mayor of Nagasaki
Hamae Okura	Chairman, Association of Women's Organizations
Toshiko Kanda	Secretary General, Consumers Japan
Koji Sasaki	Chairman, Japan Chain Stores Association
Kuniko Takahashi	Professor, Faculty of Education, Gunma University
Teiji Nakamura	Chairman, The Japan Dietetic Association
Yukio Hattori	Principal, Hattori Nutrition College
Shu Hara	President, Japan Food habitat improvement promoters association
Chieko Fukushi	Deputy General Manager, Life Information Department, Yomiuri Shimbun Tokyo Head Office
Yoshiaki Itsumi	Vice President, Parents & Teachers Association of Japan
Sho Watanabe	Board Chairman, National Institute of Health and Nutrition

many ministers who played a central role in the country participated in the sessions. Apart from the discussions at the committee, opposing views began to emerge, mainly from the opposition parties. The objection was that people should be free to decide what kind of food they want to eat and that the state should not interfere in the family dining room.

On the last day of the committee meeting, the chairperson, Prime Minister Junichiro Koizumi, in his closing remarks, referred to the controversy between the army and the navy over beriberi, and at the end of the meeting said,

"In Japan, the nation once made a mistake in its nutrition policy, causing many deaths and inconveniencing people. Nutrition is important for the nation, and I want to make this law the basic law of the nation". He was sitting right in front of me, and I remember being moved by the fact that he was not an ordinary person.

3.3 Establishment of the National Institute of Nutrition and the Rice Riots

3.3.1 Dr. Tadasu Saiki and Nutrition

One of the most important people who contributed to the development and spread of nutrition in Japan was Dr. Tadasu Saiki. He studied physiology, biochemistry, and also bacteriology at Yale University in the United States in 1905 (Meiji 38). His research work was going well and he had planned to settle in the United States, but when he returned to his home town of Saijo in Ehime Prefecture to see his grandfather who had cancer, he found out about the current situation of nutrition in Japan and decided to pursue his research work in Japan. In 1913 (Taisho 2), he came to Tokyo and became the director of Kanasugi Internal Medicine Sanatorium in Surugadai, Kanda, where he treated patients and conducted laboratory research.

At that time, nutrition science was just beginning to understand the need for trace amounts of vitamins and minerals in addition to the nutrients that provide energy, and the main task of research was to discover new nutrients. In other words, many nutritionists were immersed in basic research, and practical and applied research in nutrition was not considered to be a scientific subject. This tendency is still somewhat in evidence today, but nutritionists at that time were not interested in real-world nutritional problems.

However, Tadasu Saiki had a strong interest in the practical aspects of nutrition, because he believed that nutrition was meaningful only when it was put into practice to help people. He established the "Nutrition Research Institute" with his own funds in 1914 (Taisho 3) in Sanko-cho, Shirokane, Shiba, Tokyo. At the Institute, various research on nutrition was conducted, and terms such as "eclipse", "nutritional diet", "complete nutritional diet", "nutritional efficiency", "nutritional guidance", and others which are now commonly used in the field of nutrition, were born at the Institute. In 1920 (Taisho 9), with the aim of actively promoting nutritional policy, the government make Saiki's Institute a National Institute. The following year, a new building was established in Koishikawakagomachi, and the first director was Dr. Tadasu Saiki.

3.3.2   Establishment of the National Institute of Nutrition

Why did the Japanese government decide to take an active role in nutrition policy by establishing a national nutrition research institute? It was due to the serious historical background of the time. The specific trigger was the "rice riot" that occurred in 1918 (Taisho 7). With the modernization of the nation, the income of the people increased, and people were able to eat rice. In the case of farmers, their income also increased through sericulture and other means, and they were able to eat rice instead of wheat and Japanese millet. However, the economic boom caused by World War I led to an increase in the number of industrial workers and an exodus of human resources from the farming sector, resulting in sluggish growth in rice production. Furthermore, rice imports decreased due to the effects of the war, and the price of rice gradually became higher, and landowners and merchants began to speculate in rice and to trade in rice. The price of rice soared.

3.3.3   Rice Riots and Nutritional Research

Under these circumstances, the Terauchi Cabinet announced the invasion of Siberia as a foreign policy on August 2, 1918, and distributors and merchants further accelerated their efforts to stock rice in order to take advantage of the special demand caused by the war. The price of rice skyrocketed abnormally, making it difficult for the common people to obtain rice, an anti-government movement to "give us rice" broke out in major cities, and rallies and incidents of arson took place all over the country. More than one million people participated in the movement, which developed into a nationwide riot. This was the "Rice Riots".

The government realized the importance of nutrition and established the "National Institute of Nutrition". At the institute, comprehensive research on nutrition, including physiology, pathology, bacteriology and chemistry, was carried out, just as Dr. Saiki had dreamed. On the other hand, the Institute was also actively engaged in activities to promote nutrition through newspapers, radio and general magazines. Although a great deal of information on nutrition was distributed in the dissemination activities for improved nutrition, the underlying idea was to increase protein, fat, vitamins, and minerals by enriching the content of side dishes, rather than relying too heavily on rice as the staple food. This had the public health goal of making people healthier through improved nutrition, as well as the policy goal of alleviating the people's dependence on rice and preventing a recurrence of the rice riots. Dr. Saiki advocated for the "Economic Nutrition Act" because he believed that, in the midst of poverty, there was a need for improved nutrition through inexpensive food, and at the same time, improved nutrition would lead to an increased labor force and economic development.

3.4   The State of Nutrition During and After World War II and the Birth of the Nutritionist System

Dr. Saiki felt that with the serious food situation and the low knowledge level of the people at that time, it would be difficult to improve the nutritional status of the people only by providing information by nutrition researchers through the mass media. In other words, he thought about training practical nutrition instructors.

It was difficult for food preparers to study medical topics so Dr. Saiki tried to create a profession that had knowledge of both medicine and food preparation. This situation was also occurring on a global scale. In the twentieth century, advances in nutritional science led to the discovery of various nutritional deficiencies, and the effectiveness of dietary improvements in their prevention and treatment became apparent. Specialists began to emerge to disseminate and educate people about these results. The malnutrition caused by food shortages during World War I was particularly serious, and experts who provided guidance on improving diets based on nutritional science were highly regarded by society. As a result, activities to improve nutrition, which had initially been undertaken as a hobby or by volunteers, grew into the profession of nutritionist.

3.4.1   Establishment of a Nutrition School

In 1924 (Taisho 13), at the site of a private nutrition research institute, Dr. Saiki established Japan's first "Nutrition School" to train nutrition professionals (Photo 3.1). The school's faculty included leading nutrition researchers of the time. At that time, there were no textbooks, so the content of the education was based on practical application. In 1926 (Taisho 15), the first 13 graduates of the school completed their studies, and they were called "dietitians" (Photo 3.2). Graduates went on to work as food researchers, nutrition specialists in government agencies, hospitals, and food service facilities, and became active on the front lines of nutrition improvement. However, being a dietitian was not yet an official specialty recognized nationally.

Photo 3.1   Saiki Nutrition School at the foundation (around 1924)

Photo 3.2 First graduation ceremony of Saiki Nutrition School (March 15, 1926, Tokyo Shiba Kanasugi Kawaguchichō). In circular inset – Dr. Tadasu Saiki

In 1945 (Showa 20), the government enacted the "Dietitian Regulations" in Ordinance No. 14 of the Ministry of Health and Welfare. The government had understood the importance of nutrition even before the war, and because malnutrition had become a serious social problem due to the war, a proactive nutrition policy became essential. During the confusion, not to mention the many deaths from starvation, disreputable nutritional foods and health methods became popular, and people were unsure of what to believe and what to eat. The government enacted the "Dietitian Regulations" with the following objectives in mind: (1) to establish the status and duties of dietitians on a national basis, and to unify and provide thorough guidance on national nutrition and (2) to strengthen nutritional guidance after the provision of food for factories, business establishments, and rural areas, which were the foundation of the war effort.

3.4.2   Founding of the Japan Dietetic Association

For the wartime government, improving nutrition was an important national issue, and the training of dietitians was an urgent task. The "Dietitian Regulations" stated that dietitians were "those who use their expertise to provide nutritional guidance to the public" On May 21 (1945), the year of the end of the war, a general meeting was held at the Imperial Hotel to establish the "Great Japan Dietetic Association" in the midst of the air raids on Tokyo. In fact, the ceremony held in 2009 (Heisei 21) to commemorate the 50th anniversary of the founding of the Japan Dietetic Association was also held at the Imperial Hotel as the sacred place where the Dietetic Association had its origin. After the end of the war, it was decided to make the event even more grand, and the "The 1st General Meeting of the Japan Dietetic Association" was held at the Takarazuka Theater on October 21 and 22, 1946 (Showa 21) (Photo 3.3). The ceremony was a great success, with the participation of Dr. Tadasu Saiki and others

Photo 3.3 The First General Meeting of the Japanese Dietetic Association (at the Takarazuka Theater)

who had worked hard to establish the dietitian system, Dr. Crawford Sams from Public Health and Welfare in GHQ, Director Yukimasa Miki from the Ministry of Health and Welfare, and the governors of Osaka and Hyogo.

The following year, 1947 (Showa 22), the "Dietitian Regulations" became the "Dietitian Act" with Law No. 245, which came into effect in January 1948 (Showa 23). Dietitians became a national qualification in name and reality. At that time, 18 schools were approved as training schools, including Saiki Nutrition School, Women's Nutrition Junior College, Japan Women's University, and other women's colleges and vocational schools.

Improving nutrition during and after the war, when food was scarce, was extremely difficult. Many people in the field of nutrition bravely tackled this difficult task. We can learn about some of these efforts from the experiences of Setsuko Honda, who worked as a dietitian at the time, in her book "Living on Food: Hon no Izumi-sha". She studied nutrition at the Shokuryo Gakko (Food School) in 1945 (Showa 20). Her practical training while at the school included menu planning, cooking, and serving, and her research project was to gather potato vines, crayfish, and snails and develop a cooking for them. She was also involved in the development of "Special Pan", a bread invented by the Imperial Japanese Army to conserve flour and enhance nutrition. In addition to wheat flour, soybean flour, fishmeal, carrots, and spinach were kneaded into the dough of this bread to make it more nutritious. After graduation, she was recruited as an army dietitian at the Oita Men's Flying Corps School, where she was in charge of the meals for about 1000 soldiers. Around June of the last year of the war, "kamikaze pilot" who was ready to die came to her and said, "I'm leaving for Okinawa tomorrow", to thank her for his daily meals, and she was moved to tears.

3.4.3   The End of the War and the Deterioration of Nutritional Conditions

Immediately after the war, there was not enough food to maintain the lives of all the people. Around August 1945 (Showa 20), Tokyo was burnt to the ground, the suffering from hunger reached a peak, the black market was rampant, and many people turned to black-market rice to survive. Since the sale of rice was controlled by the government, purchasing rice from the black market was illegal. However, people could not meet their nutritional needs through rationing alone, so many people turned to the black market. In 1947 (Showa 22), a shocking incident occurred: Judge Yoshitada Yamaguchi of the Tokyo District Court starved to death because he refused to eat black market rice out of a sense of justice as a law-abiding citizen. This became big news.

According to the data published by the Society for the Study of Living Issues in 1946 (Showa 21), the amount of daily nutrition obtained from rations was 1209 kcal and 32.2 g of protein, while that from non-rations was 765 kcal and 26.9 g of protein.

There was no way for the common people to survive without breaking the law. However, both the government and GHQ strengthened the crackdown under the Food Control Law in order to preserve law and order in the country. For example, in 1948 (Showa 23), when the crackdown was most severe, the number of arrests nationwide reached 917,324 cases and the number of arrestees 927,301 people. The food shortage became a social problem, demonstrations of "Give us rice" took place all over the country, and the worsening of nutritional conditions due to food shortages became an important issue for the nation.

3.5 Postwar Reconstruction and Improvement of Nutrition in Japan

3.5.1 Setting Up the Nutrition Improvement Act

At the same time that it was working to disseminate nutritional knowledge through newspapers and radio, the government formulated the "Nutrition Improvement Act" in 1952 to make it mandatory to assign dietitians and to facilitate the implementation of nutritional improvements.

The purpose of this law was to assign dietitians to group meal facilities to provide meals with excellent nutritional balance while utilizing the limited amount of food available, and to ensure nutrition education. Using facilities that serve meals to a large number of people at the same time, such as schools, industrial concerns, hospitals, and welfare facilities, dietitians began to improve the content of the meals served and to spread knowledge of nutrition, and this work was called "nutrition guidance". The speciality of the dietitian was positioned as "nutritional guidance". This distinctive Japanese method of improving nutrition, which combines the provision of meals with nutritional education, is rare world-wide, and we believe it is a method to be proud of.

3.5.2 This "Kitchen Car" Is Not a "Food Truck"

Immediately after World War II, the Japanese suffered from severe malnutrition with no food or money. Initially, the country relied on food imported from the United States. On the other hand, in the victorious United States, advances in agricultural technology had led to an excess of crops. The US government considered selling this excess food to foreign countries, and one target was Japan, which was suffering from food shortages. In 1954, a meeting was held between Japan and the United States on "the use of market development costs associated with the acceptance of surplus agricultural products in the United States." However, there was no money in Japan to buy these products, and it was difficult for Japanese people who ate rice as their staple food to accept wheat bread and dairy products.

After discussions, the decision was made on the condition that the US government could hold the money that Japan to purchase food from American farmers for a while and use part of the money to promote and spread imported food in Japan. In recent years, some have argued that this was a long-term strategy to westernize the Japanese diet and make it dependent on American agricultural products. However, the only way to survive the severe hunger of the time was to rely on imported food must be from abroad. In particular, supplementing nutritionally inadequate Japanese food with highly nutritious Western food was an effective means of resolving malnutrition. Furthermore, the important point is that advertising funds were provided to promote unfamiliar imported food among the Japanese. Japan actually provided nutritional education to resolve malnutrition, although it was said that this was an advertising expense for the United States.

The Japanese government with the money for advertising expenses, bought a "Kitchen Car" in which the rear part was converted into a kitchen for cooking demonstration, and dietitians boarded and provided nutrition education to every corner of Japan. This "kitchen car" was not a "Food Truck" that provides meals as can be seen in the United States today, but a special car that could provide nutrition education that also served as a cooking demonstration (Photos 3.4 and 3.5). Dietitians along with Health Mates who were dietary habits improvement promoters as volunteers participated in the community nutrition improvement.

Photo 3.4 Kitchen Car on Japan in 1950

Photo 3.5 American!s Food Truck in 2020

This sort of activity has become widely recognized in Japanese society, and dietitians have been assigned to health centers, schools, companies, hospitals, long-term care facilities, and so on. This has allowed Japan to create a social environment in which people can access a healthy and nutritious diet wherever they eat. In other words, "Japan changed the US food policy into a nutrition policy that made people healthier", and this result has since become a deterrent to obesity in a society where the economy is highly successful and food is abundant.

3.5.3 Improved Nutritional Status

With the improvement of the food situation and the nutritional guidance by dietitians, malnutrition in the postwar period was rapidly and peacefully addressed. In order to understand the results of the improvement of nutrition in Japan, it is helpful to refer to the "Trends in Nutrition in the Postwar Showa Period: Looking Back on

40 Years of the National Nutrition Survey" supervised by the Nutrition Guidance Institute of the Japan Dietetic Association. The basic data for this report came from the "National Nutrition Survey" conducted in 1945 (Showa 20) by order of the GHQ in order to establish the basis for emergency food measures.

The "National Nutrition Survey" was later included in the "Nutrition Improvement Act" and became a basic factor in the PDCA cycle (plan-do-check-act cycle) for understanding and improving the nutrition of the Japanese people. "The National Nutrition Survey" was renamed the "National Health and Nutrition Survey" in 2003 (Heisei 15) and is conducted annually to provide basic data on the nutritional status of the Japanese people and their nutrition-related health conditions. Based on these results, problems are identified and a plan is drawn up to solve them. Following this plan, dietitians take the lead in providing nutritional guidance to the public. No other nation has implemented such a thoughtful nutrition policy, and this is the basis of "Japan Nutrition".

In the report, the postwar recovery of nutrition is classified into four periods.

① Immediately after the war to 1948 (Showa 23): Postwar turmoil

A period of severe hunger and malnutrition caused by extreme food shortages, with many starvation deaths in the cities.

② From 1949 to 1954 (Showa 24~29): A period when the food situation gradually improved

School lunches became a full meal program in 1950 (Showa 25), and the nutritional status of children improved. This was an era of increased intake of animal products, legumes, fats and oils, and of a marked increase in the intake of animal protein, fat, calcium and vitamin A.

③ The 1950s (Showa 30): A period of rising national income known as the "consumer revolution"

During this time, consumption of ham, sausage, and instant noodles increased, and dietary habits began to become more Westernized and diversified. The consumption of fats and oils, meat, eggs, milk and dairy products also increased, and the intake of protein, vitamins and minerals also increased, leading to an average improvement in the nutritional status of the Japanese people.

④ 1960s (Showa 40): a period of high economic growth that led to a significant rise in incomes

Except for carbohydrates, intake of all nutrients increased.

⑤ From 1975 to 1988 (Showa 50~63): The mild decline in rice consumption continued with other food groups remaining unchanged

The postwar problem of low nutrition had been completely solved and nutrition status became relatively stable (Fig. 3.1).

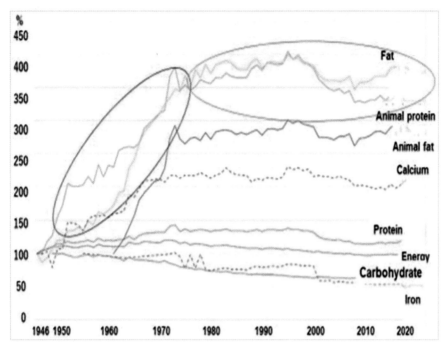

Fig. 3.1 Trends in nutrient intakes
Note: Nutrient intakes in 1946 set at 100 (Carbohydrate intake in 1949, animal fat in 1970 and iron in 1972 set at 100)

3.6 Achievements in Improving Nutrition and the Crisis of the Dietitian System

As nutritional improvement progressed and people were able to break free from nutritional deficiencies, they began to forget how grateful they had been to be able to eat and the significance of eating with nutrition in mind. This led to the emergence of the theory that nutrition and dietitians were unnecessary. At the root of this was the belief, held by many people at the time, that "nutritional problems would be solved by economic development". People thought that nutrition was important because of the extreme poverty and food shortages caused by the war; however now that the war was over and the economy had developed, they thought that the "nutrition problem was solved". This is a natural consequence for people who have suffered from hunger and nutritional deficiency caused by long-term poverty. They could not have imagined that the time would come when there would be an abundance of food and they would suffer from overnutrition.

3.6.1 Prevention of the Bill to Abolish Dietitian Act and the Establishment of the Society for Nutrition Improvement

In 1951 (Showa 26), a debate arose in the Local System Council to abolish the Dietitian Act. It was argued that nutrition no longer needed to be a national policy. The Japan Dietetic Association campaigned to prevent the repeal bill on the grounds that there was still insufficient improvement in nutrition for the poor and rural areas. This was the first time that The Japan Dietetic Association waved a flag against the government. At the same time, there was some concern that the status of dietitians would be unstable and that they would not be able to take root as a distinct profession under the "Dietitians Act" alone. In other words, there was debate about the need for a system that would legally support the status of dietitians. As a result, in 1952 (Showa 27), the "Nutrition Improvement Act" was promulgated and enforced with the aim of improving the health and physical strength of the people. This Act included a national nutrition survey, nutritional consultation offices, specialized nutritional guidance by prefectures, a system of nutritional advisors, nutritional management at group meal facilities, special nutritional foods, and nutritional labeling. The Act also instructed the public on the need for improved nutrition and required that dietitians be assigned to group meal facilities. Dietitians were now recognized both in name and reality as an official profession. Incidentally, "The Japan Society for the Improvement of Nutrition" was established in 1954 to take advantage of this formulation to study the significance and methods of improving nutrition in an academic setting.

In 1957 (Showa 32), another unpredicted crisis for dietitians occurred. This was the introduction of the "Cooking Improvement Bill" in the Diet. The proposal was made by a group of cooks, and they demanded a law that aimed to make the placement of cooks mandatory, just like dietitians. An opposition alliance was formed involving the Japan Dietetic Association, Saiki, and the Medical Association. The reason for the opposition was that "the purpose of improving cooking was already included in the Nutrition Improvement Act, and there could be no improvement in cooking without a foundation in nutrition, and this was a job that could only be done by dietitians". About 70,000 opposition signatures were gathered, and in the end, the "Cooking Improvement Bill" was defeated at the last minute, and the authority regarding menu planning and cooking for the purpose of improving nutrition as the work of dietitians was upheld.

The nutrition improvement movement developed into a national movement, and, combined with postwar reconstruction, it contributed greatly to improving the nutritional status of the Japanese people. Nutritional guidance, which enabled people to make effective use of limited food resources, was widely accepted by society, and the social reputation of dietitians gradually increased. Dietitians prepared menus and provided nutritional guidance for group meals at factories, offices, schools, and other facilities, and also held seminars nation-wide and provided nutritional education using the mass media.

3.7 Emergence of Lifestyle-Related Diseases and the Birth of the Registered Dietitian system

With the promotion of nutritional guidance, the increase in agricultural products, and the development of the economy and distribution, the problem of low nutrition caused by food shortages was almost completely solved by the 1960s. The adverse effects of the Westernization of the diet began to occur, and obesity and non-communicable diseases: NCDs, also known as lifestyle-related diseases, began to increase. nutritional deficiency diseases caused by food shortages result from unbalanced foods, improper cooking, and other factors for which solutions can be found in agricultural science and home economics. NCDs are caused by individual habits and metabolic disorders, and solutions could not be found without the development of clinical nutrition research and education.

In addition to the training of dietitians by junior colleges and vocational schools, the need to train high-level dietitians at universities became apparent. In April 1962 (Showa 37), the Dietetic Council of the House of Councilors Committee on Social Welfare and Labour proposed a "New Dietitian System". According to the records of the Council, "In the past, dietitians have been responsible for the nutrition, rational consumption, and nutritional effects of food in group meal facilities, but for complex and difficult cases, dietitians with special training are required".

In September of that year, the Diet approved an amendment to the "New Dietitians Act" to establish a system of registered dietitians to combat NCDs. In April 1962, the Department of Nutrition was established at the Faculty of Medicine of the National University of Tokushima. Dr. Keizo Kodama, then dean of the Faculty of Medicine at the University of Tokyo, became president of the University of Tokushima and established the Department of Nutrition at the medical school to research and teach nutrition as part of medical education.

In April 1963 (Showa 38), the Dietetic Council submitted a proposal to the Minister of Health and Welfare for the "establishment of specialized departments and divisions to enable students to obtain the title of Registered Dietitian: RD" when it reported on the "Standards for the Examination and Training Institutions for registered dietitian". In response to this proposal, the Ministry of Education approved a course for the training of RDs at the Department of Nutrition, Faculty of Medicine, National University of Tokushima, and appropriated funds for the expansion of the course in the government budget for 1964 (Showa 39).

The substance of the Council's conclusions at that time was as follows.

1. Concept of the new Department of Nutrition

In considering the criteria for a new Department of Nutrition to award the title of Bachelor of Nutrition, it is appropriate to adopt the following policy:

(a) This Department of Nutrition must have enough content to become a stand-alone department in order to be able to award a new bachelor's degree even if it is placed within an existing department.

(b) This Department of Nutrition shall target a new comprehensive field of academic research on nutrition and shall be distinguishable in content from the existing Department of Nutrition in the Faculty of Agriculture and the Faculty of Home Economics.
(c) The main purpose of this curriculum of the Department of Nutrition should be to train academic researchers on nutrition.

2. Characteristics of the School of Nutrition, Department of Nutrition

The new Department of Nutrition will focus on research and teaching in the basic areas of nutrition, and food preparation may not be required.

As a result of the above, the Standards for the Establishment of Universities were established in March 1965 (Showa 40) with Ordinance No. 7 of the Ministry of Education, and the system of RD was officially born.

Although it was decided to train a registered dietitian as a professional occupation for NCD measures, the specific work was not fully defined, and the law stipulated that an RD was a "person who performs complicated and difficult work." For the past 35 years, the specific duties of RD have remained unclear, and also the division of roles between dietitians and RD has remained unclear until 2000.

3.7.1   Pathological Nutrition Workshop and End

Under these circumstances, the National Training Seminar on Pathological Nutrition was held by the Japan Dietetic Association in 1971 (Showa 46). The textbook "Pathological Nutrition" (Dai-ichi Shuppan) (Photo 3.6) was produced by prominent clinicians. It was actually planned and edited by Kiku Morikawa, the third president of the Japan Dietetic Association. He (Photo 3.7) was one of the first to recognize that RD in Europe and the United States were trained as part of the same medical professions as doctors, nurses, and pharmacists, and he had a strong desire to make RD in Japan similar.

At this time this training program was equal to the best in Europe and the United States. Many aspiring dietitians from all over Japan took part in this training program, and it became a pioneering program. However, the problem with this training program was that it was conducted without determining the purpose of the training, the status of those who completed the training, or the duties of those in the medical field. The training program ended in 1988 (Showa 63) without developing into a specific qualification system. In the end, the training ended with dietitians only learning the basic knowledge of medicine to teach medical students.

Photo 3.6   Textbook of Pathological Nutrition Workshop

Photo 3.7   Kiku Morikawa, the third president of the Japan Dietetic Association

But in 1978 (Showa 53), the 5 points for nutritional dietary guidance by a registered dietitian as a medical fee covered by universal health insurance was newly added for chronic disease guidance. At the time, it was said that the 50 yen (50 cents US) fee was not even enough to pay for a cup of coffee, but this was the first time that a registered dietitian who was not a doctor had received such a fee, and it was the first time that a dietitian's professional skills were recognized. Since then, it has become, in 2020, 260 points for 30 min for the first consultation and 200 points for the after the second time.

In 1982 (Showa 57), another major event occurred. As part of its policy of administrative simplification, the government considered a proposal to abolish the "Dietitian Act". The government believed that nutritional deficiencies had been eliminated, and that although a dietitian system had been created, its role was unclear, and the government no longer needed to be proactive in its nutrition policy. This was a third crisis for dietitians. The Japan Dietetic Association was at the forefront of the campaign to stop the abolition of the "Dietitian Act", collecting petitions from the public to withdraw the bill to abolish dietitians, and staging demonstrations at the Diet. The reason for the opposition was not that there were no more nutritional problems in our country, but rather that "non-communicable chronic diseases: NCDs caused by overeating have increased, the nutritional problems of the people have diversified, and nutrition policy is becoming more important". The government promised to work actively on the prevention of NCDs, and the proposal to abolish the dietitian system was eventually scrapped (Photo 3.8).

Photo 3.8 Blocking the proposed abolition of the Dietitian system

The public movement against the abolition of the "Dietitian Act", while raising the fundamental question of whether or not dietitians are necessary, also served as an opportunity to reaffirm the importance of dietitians, and became the catalyst for the future "Law Reform 2000".

3.8 School Lunches and Nutrition Education

3.8.1 History of School Lunches

School lunches began in 1889 (Meiji 22) as a relief measure for needy students at Chuai Elementary School in Tsuruoka, Yamagata Prefecture. Later, they were implemented in Hiroshima, Akita, Iwate, Shizuoka, and parts of Okayama prefectures. School lunches were born out of social necessity, much like today's "children's cafeterias". In 1914 (Taisho 3), Dr. Tadasu Saiki recognized the need for school lunches, and with the support of a scientific research grant from the Ministry of Education, he began to provide school lunches to children in nearby schools. In 1923 (Taisho12), school lunches were encouraged in a notice from the Vice-Minister of Education, "On the hygiene of elementary school children". In 1932 (Showa 7), the Ministry of Education issued a directive entitled "Temporary School Meal Service Facilities", and school meals were provided to poor children with government subsidies. In 1940 (Showa 15), school lunches were extended to include not only poor children but also malnourished children and physically weak children, and in 1944 (Showa 19), school lunches were provided to about 2 million elementary school children in six major cities with special rations of rice, miso and other foods. In other words, school lunches had already been implemented before the war as part of the campaign to improve nutrition.

The present school lunch, in which all students eat the same meal together, did not start as a national policy until after the war. By the way, it is said that the resumption of school lunches after the war started with skimmed milk powder from LARA supplies, but this is not necessarily true. LARA (Licensed Agencies for Relief in Asia) was an aid organization for Asian countries established and authorized by the government's Relief and Control Commission in 1946 (Showa 21). However, there was resistance to aid to Japan. At the time, the anti-Japanese movement was still raging in the US, and there was feeling that there was no need to save the children of enemy countries. It is said that sympathetic Americans and Japanese-Americans who were concerned about their homeland desperately collected supplies. On August 30, 1946 (Showa 21), the GHQ issued a memorandum regarding the receipt and distribution of relief supplies to LARA, and the Ministry of Health and Welfare responded with a plan for distribution on September 20.

The plan stated that "relief supplies will be distributed fairly on the basis of need, without regard to nationality, religion, political party, or political faction", and that priority would be given to facilities for the socially vulnerable. This shows the excellence of the administrative officials of the time. This philosophy led to the extension of relief supplies to schools. Aid was provided from November 1946 to June 1952 (Showa 27), and totaled 33 million pounds. This included 25.22 million pounds of food, including a wide variety of nutritious foods such as whole milk, skimmed milk powder, sugar, baby food, dried fruit, soybeans, dried eggs, canned goods, and flour.

It was in the summer of 1946 that former president Herbert Hoover of the United Nations Relief and Rehabilitation Commission visited Japan and advised GHQ to resume school lunches. In October of the same year, Colonel Sams of GHQ recommended that the government implement the program, and GHQ promised to assist. However, despite the promise, there was no food for school lunches left anywhere in the country. Therefore, the canned food and LARA supplies that the former Japanese army had were used. In December 1947 (Showa 22), the Ministry of Education, the Ministry of Agriculture and Forestry, and the Ministry of Health and Welfare issued a notice from the vice-minister stating that "from the standpoint of improving the physical condition of schoolchildren and nutritional education, it is desirable to widely provide school lunches", and school lunches were officially resumed. This notice stated that school lunches should be provided with the goal of improving the physical condition of schoolchildren and nutritional education, and this principle has been carried over as the philosophy of school lunches to this day.

When school lunches were first reintroduced, food was scarce and there were no school lunch facilities because the schools themselves had been destroyed. At first, military canned foods such as corned beef, canned spinach, and tomato ketchup were used, and at one school, only tomato ketchup was served, which was sometimes eaten with a bowl of rice. The contents of school lunches gradually improved with the release of LARA supplies, but staple foods could not be provided, so families brought in boiled potatoes and bread. In 1949 (Showa 24), UNICEF began to provide support, milk and flour were distributed, and the prototype of the current school lunch with bread was created. At that time, a comparison was made between schools that received UNICEF support and those that did not, and it was reported that in six months the height and weight of schoolchildren in the supported schools increased by one year's growth compared with those in the control schools. In 1954 (Showa 29), the School Meal Law was formally enacted to include elementary and junior high school students (Photo 3.9).

Photo 3.9 Improvement of physique through school meals
The children's physique has improved remarkably as a result of school meals. Top: immediately after the start of the school meal, middle: 4 months later, bottom: 2 years later the same number indicates the same child

3.8.2   Enactment of the Nutrition Teacher System and the Nutrition Education Basic Act

Since then, reducing or abolishing school lunches has been repeatedly debated, but thanks to the efforts of those involved in nutrition, they have developed into what they are today, and are highly regarded world-wide. The reason for this is that school lunches in Japan began as a way to rescue children from hunger and poverty, but the provision of meals was positioned as part of nutrition education, and menus were considered a living educational medium (Table 3.6). If children continue to eat nutritionally balanced meals for 6 years or more during their growth period, they can physically learn what kinds of meals are preferable.

. In addition, by delivering the "School Lunch Report" to the families every week to convey the children's experiences to the families, the children can use the report as a topic to talk about nutrition at the family dining table, which improves the meals at home. With the development of the economy, the diet became richer and some of it became Westernized, but the over-all diet of the Japanese did not become Westernized, and this helped to form a Japanese style diet with an excellent nutritional balance centered on rice. This philosophy led to the development of the "nutrition teacher system" in 2005 (Heisei 17).

The nutrition teacher is a category of teacher that was established in 2005 (Heisei 17) to ensure the healthy development of children. The nutrition teacher's duties include the management of school lunches as well as the education of children so that they can develop the ability to manage their own food and acquire desirable eating habits, using school lunches as a living teaching tool. With the increase in lifestyle-related diseases, there was a need to strengthen nutrition education for children. Dietitians have long had a strong desire to actively implement nutrition education as regular teachers, and the founder of the Japanese Association for Dietetic Research and Education, President Nobu Tanaka, has long been a leader in this movement. When the Ministry of Education, Culture, Sports, Science and

Table 3.6 Goals of school lunches

1	To promote the maintenance of health through the intake of appropriate nutrition.
2	To deepen the correct understanding of meals in daily life, to cultivate the ability to make sound judgments about eating habits, and to develop desirable eating habits.
3	To enrich school life and to cultivate a cheerful sociability and a spirit of cooperation.
4	To deepen understanding of the fact that diet is based on the benefits of nature, and to cultivate a spirit of respect for life and nature and an attitude that contributes to the preservation of the environment.
5	To deepen understanding of the fact that life is supported by the various activities of people involved in food, and to develop an attitude of respect for their work.
6	To deepen understanding of the excellent traditional food culture of our country and of each region.
7	To lead to a correct understanding of the production, distribution and consumption of food.

Technology (MEXT) held a meeting of the committee to study the issue, President Tanaka told me.

"Dr. Nakamura and I are the only two who understand the necessity of this system, and all the other committee members are against it."

In recent years, Japan's school lunch system has been gaining recognition overseas. Some countries have opened outlets and restaurants in schools, or distributed commercial lunches, with the aim of eliminating food loss. However, these are school lunches that do not consider the meaning of providing meals in schools, which are educational institutions, and they are not true school lunches. The primary goal of school lunches is to function as education in nutrition, food and diet.

Timeline of the History of Nutritional Improvement in Japan

Japanese calendar	A.D	Event in Japan
Meiji 1	1868	Meiji Restoration
4	1871	Theodore Hoffmann, (German) military physician (internal medicine) introduced nutrition to Japan
5	1872	Beginning of providing meals to 300 people at the Tomioka Silk Mill in Gunma Prefecture
17	1884	Dr. Kanehiro Takagi added wheat to military rations to prevent beriberi
19	1886	Dr. Rintaro Mori wrote "The Theory of Japanese Military Food".
22	1889	School lunch started at Yamagata Chuai Elementary School
43	1910	Dr. Umetaro Suzuki discovered oryzanine in rice bran
Taisho 3	1914	Dr. Tadasu Saeki established the Nutrition Research Institute
7	1918	Request for unification under "Nutrition"
9	1920	National Institute of Nutrition of the Ministry of the Interior established
13	1924	The Extraordinary Committee on Beriberi Disease concludes beriberi is mainly caused by vitamin deficiency
		Food and Nutrition Research Institute, Faculty of Medicine, Keio University is established
		Saiki Nutrition School opened
15	1926	First graduates of the Saiki Nutrition School, 15 "Dietitians" are born
Showa 9	1934	First issue of "Journal of the Dietitians Association" published "Nutrition Society of Japan" was recognized as the 13th subcommittee of the Japanese Medical Association
13	1938	Opening of nutritional therapy clinic attached to the National Institute of Nutrition The Institute was transferred from the Ministry of Home Affairs to the Ministry of Health.

(continued)

Japanese calendar	A.D	Event in Japan
20	1945	End of war Promulgation of the Dietitian Regulations. Establishment of the All-Japan Dietetic Association
21	1946	Colonel Howe takes up his post
		National Nutrition Survey started
		Memorandum regarding GHQ LARA supplies, 1st Japan Dietetic Association held (Takarazuka Theatre).
		"The Council for National Food and Nutrition Policy" is established at the Headquarters for Economic Stabilization, Nutrition Section established in the Public Health Bureau of the Ministry of Health and Welfare.
22	1947	Start of the school lunch system
		"Dietitian Act" established to define and regulate the work of dietitians, Public Health Service Act established to assign dietitian to provide public nutrition services
23	1948	Japan's hospitals are rated as medieval by the U.S. military with the promulgation of the "Medical Care Act".
24	1949	First National Examination for Dietitians
25	1950	Inauguration of the complete food service system in hospitals
26	1951	Victory in the campaign to prevent the abolition of the "Dietitians Act"
27	1952	Promulgation of the "Nutrition Improvement Act"
29	1954	Promulgation of the "School Lunch Act"
		The Japanese Society for the Improvement of Nutrition established
33	1958	The Japan Dietetic Association's official journal "Nutrition Japan" launched
34	1959	The Japan Dietetic Association established as a corporation
37	1962	Registered Dietitian System established
		Department of Nutrition, Faculty of Medicine, The University of Tokushima founded
46	1971	Japan Dietetic Association's "Course on Pathological Nutrition Techniques" started
53	1978	Dietician's nutritional guidance added to medical service fees
57	1982	Promoting the campaign against the abolition of the dietetic licensing system
60	1985	The National Examination System for Registered Dietitians established
		Inauguration of the Japan Society for Venous and Enteral Nutrition
62	1987	The Japan Dietetic Association's "Lifelong Learning System" established
Heisei 6	1994	New hospitalization meal treatment system established
9	1997	Implementation of the Community Health Act
		Dr. Norimasa Hosoya Chairperson, "Committee on the Role of Registered Dietitians in the 21st Century", Ministry of Health, Labour and Welfare

(continued)

Japanese calendar	A.D	Event in Japan
12	2000	Amendments to the "Dietitians Act", changing from registration to licensing of dietitians
14	2002	The "Nutrition Improvement Act" becomes the "Health Promotion Act"
17	2005	Nutrition management teams added to medical services
18	2006	Promulgation of the "Nutrition Education Basic Act"
20	2008	Nutritional administration added to medical service fees, "Nutrition Japan" becomes "Journal of the Japan Dietetic Association", Specific medical consultation and specific health guidance started, Nutrition Care Station opened The 15th International Congress of Dietitians (ICD2008) is held.
22	2010	Nutritional support teams added to medical service fees
23	2011	The Great East Japan Earthquake, JDA-DAT established
24	2012	The Japan Dietetic Association established as a public interest incorporated association Abolition of the additional fee for nutritional administration; included in the basic hospitalization fee calculation requirements
26	2014	The Japan Dietetic Association's "Lifelong Learning System" becomes the "Lifelong Education System".
28	2016	Establishment of Nutrition Day (4 August) and Nutrition Week (1–7 August)
30	2018	Certified Nutrition Care Station Program Launched Simultaneous revision of medical and long term care fees in fiscal 2018, provision of nutritional information is rated.

Bibliography

Fujisawa Y (2000) Dietitians and nutritionists are professional occupations that support the 21st century. In: Dietitians and nutritionists complete guide. Footwork Publications, pp 8–14

Fujiwara H (1960) The history of nutrition in Japan (Part 2). Shokuseikatsu 54(4):61–86

Hayano T (2019) The roots of the "Dietitian Act" and tomorrow's dietitians and nutritionists. J Jpn Dietetic Assoc 62:3–11

Kido H, Suzuki Y, Azuma J et al (2004) Re-validation of Kanehiro Takagi's experimental voyage by meta-analysis. Jikei Med Univ J 119:279–285

Masatoshi H (2009) Development of the Dietitian System, from the "Nutrition Improvement Act" to the "Health Promotion Law". J 50th Anniversary Incorp Jpn Dietetic Assoc:130–140

Nakamura T (2019) The first modernization of diet and nutrition. Clin Nutr 134:115–118

National Dietetic Association (ed) (1981) History of dietetics in Japan, "Dietitian Act" and "Nutrition Improvement Act", group meal by release food. Shujunsha:252–258; 213–226

Nehme AE (1980) Nutritional support of the hospitalized patients-the team concept. JAMA 243:1906–1908

Ohta M (2016) Modernization of food and nutrition. Modern Acad:117–113

Oiso T (1972) Nutritional essays. Medical and Dental Publishing

Saiki Y (1986) Nutritionist Tadasu Saiki. Gendousha

Yakuwa S (2009) Trends in the dietitian system. J 50th Anniversary Incorp Jpn Dietetic Assoc:26–57

Yamazaki I (2006) Chinese medicine nutrition. Dai-ichi Shuppan

Chapter 4
Transformation to Human Nutrition

Abstract At the time when people had solved the problem of food shortages, nutrition had lost direction with regard to future research, education, and practice in Japan. But since the 1970s, it has become apparent that some of the injured, sick, and elderly people are suffering from malnutrition. Dr. H. Norimasa argued that in order to improve human health and to prevent and treat diseases, it is necessary to clarify the internal dynamics of nutrition at the individual level, the organ and tissue level, and the cellular level, which is expressed in the comprehensive concept of "Human Nutrition".

In 1997 the Ministry of Health, Labor and Welfare established the "Committee to Study the Role of Dietitians in the 21st Century". This committee concluded that it is necessary to introduce nutrition care management to improve the nutritional status by human nutrition. A registered dietitian is a person who is licensed by the Minister of Health, Labor and Welfare and works under the title of registered dietitian to provide nutritional guidance necessary for the medical treatment of injured or sick persons, nutritional guidance for the maintenance and promotion of health that requires a high level of specialized knowledge and skills in accordance with the physical condition and nutritional status of individuals. The role of registered dietitians in medical and welfare care has changed dramatically Various types of malnutrition appeared, and this led to a decline in the quality.

Registered dietitians were placed in all fields of health care (primary prevention), medical care (secondary prevention), and welfare (tertiary prevention). The ground design for "Japan Nutrition" had been created.

© The Author(s) 2022
T. Nakamura, *Japan Nutrition*, https://doi.org/10.1007/978-981-16-6316-1_4

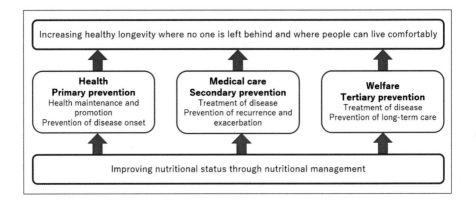

Keywords Dietary reference intakes · Committee to Study the Role of Dietitians in the twenty-first century · Clinical dietitians system · Nutrition Japan · Japanese Society of Clinical Nutrition · Law reform 2000 · Registered dietitians · Nutrition care process · American Academy of Nutrition and Dietetics · Nutrition Care Process (NCP) · Nutritional intervention · PES method

4.1   Human Nutrition and Norimasa Hosoya

At the time when we had solved the problem of food shortages and malnutrition caused by an overemphasis on staple foods, nutrition in Japan had lost direction with regard to future research, education, and practice. In other words, there were no prospects for the future. Overeating, obesity, and lifestyle-related diseases were becoming a problem due to the simplification and westernization of the diet, but these were considered to be individual lifestyle problems caused by lack of exercise and overeating, and not problems that needed to be addressed by society as a whole. By daring to call non-communicable chronic diseases "lifestyle-related diseases", it was implied that what was wrong was "your lifestyle, not a national problem". Although there were experts who complained about the harmful effects of obesity and lifestyle-related diseases, there were only a few who studied the subject head-on. There were also intellectuals who argued that nutrition and dietitians were unnecessary, and that qualification as a dietitian was becoming part of the tool set for brides.

It was often said in wedding speeches, "A groom who marries a dietitian will be lucky to have good food and good health." Some of the professors who trained dietitians even said that if all housewives were qualified as dietitians, it would be possible to improve nutrition at home, thus eliminating the need for education of professionals. At the root of the low regard for the academic value of nutrition and the professional reputation of dietitians was the belief that nutrition problems were ultimately caused by poverty and food shortages, and would be solved naturally if countries became prosperous. Indeed, the world's nutrition problems were concentrated in the developing countries of the Southern Hemisphere, where poverty was the greatest cause.

4.1.1   Diversity of Nutritional Issues

At the same time, the nutrition problem was becoming a situation that could not be solved by simple economic means alone. This is because it has become clear that, as people become more affluent, obesity due to overnutrition caused by overeating, as well as non-communicable diseases such as diabetes and cardiovascular diseases, become more frequent, which increases medical costs and affects national finances. Even in affluent Western industrialized countries, a new type of undernutrition that does not depend on food shortages is emerging among the sick, the elderly, and young women. In particular, since the 1970s, clinical malnutrition and disease-related malnutrition have become social problems in Western industrialized countries. In Japan as well, it has become apparent that some of the injured, sick, and elderly people who are admitted to hospitals and welfare facilities and who eat menus prepared by nutritionists are suffering from malnutrition. Moreover, it has become clear that if such a condition is left untreated, the effectiveness of surgery and drug therapy decreases, the level of nursing care increases, and the number of days spent in hospital increases, which ultimately increases medical and nursing care costs.

4.1.2   The Birth of Human Nutrition

Many nutrition professionals in our country were unable to find a methodology that could address the diversity and complexity of these nutritional issues. At this time, Professor Norimasa Hosoya of the University of Tokyo appeared on the scene like a savior. He was a leading researcher on the digestion and absorption of nutrients, and he argued that the issue of nutrition should be expanded from the ingestion of nutrients to the internal dynamics of the body, and that the nutrition of the human body should be comprehensively evaluated and judged using biochemical methods such as blood tests and urine tests. In other words, in order to improve human health and to prevent and treat diseases, it is necessary to clarify the internal dynamics of nutrients at the individual level, the organ and tissue level, and the cellular level, and to improve the condition of the body, which is expressed in the comprehensive concept of "Human Nutrition". Therefore, he said that the evaluation of foods and diets should not be determined merely by the nutrient content, as in the past, but by their ability to improve nutritional status. In the case of malnutrition caused by food shortages, the risk was on the food side, and the problem could be solved by improving the food supply, food selection, and menu, etc. However, in the case of obesity, lifestyle-related diseases, and malnutrition of the injured, sick, and elderly, the risk was on the human side. Dr. Hosoya explained that we should reconsider nutrition from the human side.

4.2   Initiatives for Human Nutrition

4.2.1   Encounter with Dr. Norimasa Hosoya

I first met Dr. Norimasa Hosoya at the "Regional Workshop on Nutrition Policy and Supporting Program" held at Ewha Women's University in Seoul in 1977 (Showa 52). two to three experts from each of a number of countries, mainly in Asia. the FDA, WHO, and UNICEF participated in the workshop, which was held for 2 weeks at the International House of the university, where attendees ate and slept. The purpose of the conference was "how to save children in Asia and Africa suffering from nutritional deficiency", and it was the first time for me to participate in a full-scale workshop. I had to complete a grueling schedule of keynote speeches and plenary sessions in the morning, group discussions on each theme in the afternoon, and report writing from evening to midnight. Finally, exhausted by the complexity of the discussions and the stress of the English language, I ended up urinating blood. In addition to this, when I returned to my room, Dr. Hosoya came to my room every night with a glass of whiskey in his hand, and gave me a long, one-sided talk, saying, "'Nakamura-kun , from now on, it is human nutrition."

However, to tell the truth, I did not understand this well, because I was a novice and did not know the situation of nutrition problems and nutrition research in Japan at that time.

In the first place, nutrition is a science for human beings, and I didn't think it was necessary to call it "human nutrition".

However, after about 10 years, I gradually came to understand the meaning of the term "human nutrition", which is to say that although nutrition can contribute to the maintenance and promotion of human health and to the prevention and treatment of diseases, the approach that makes this possible should be based on food and people. In other words, the conventional method of surveying diets, calculating nutrient intakes, comparing them with requirements, clarifying problems, and providing nutritional guidance is ineffective in a situation where under-eating and over-eating coexist, and where malnutrition is caused by internal factors such as disease and aging. I understood in my own way that Dr. Hosoya was arguing that we need to be closer to human beings, and that we need to construct new research, education, and practice methods for nutritional science, starting from human beings. When I told a famous scholar that it took me 10 years to understand what Dr. Hosoya was saying, he said, "That's too soon. In general, it takes about 20 years."

4.2.2 Overseas Training

In order to learn more about human nutrition, there was a time when I went to the United States and Australia every year from 1993 to 1996 with 20–30 volunteers led by Dr. Hosoya. We visited and trained at the Ohio State University, Stanford University, California State University Sacramento], the University of Minnesota, the Mayo Clinic, the University of Sydney, and so on. During this training, I learned that research and education in nutrition is [are] conducted in the biochemistry and public health departments of medical schools, and that dietitians work not in kitchens and offices but in hospital wards, where they are trained as medical professionals in the same way as doctors, nurses, and pharmacists. The dietitians in the wards were practicing and researching clinical nutrition in cooperation with other professions. At each training site, a mountain of materials was handed out and the study was hard, but it was fresh and shocking.

Dr. Hosoya and I shouted, "This is a black ship coming to the Japanese nutrition world."

4.2.3 From Nutritional Requirements to Dietary Intakes Standard

I attended an impressive symposium at the International Congress of Nutrition held in Montreal in 1997. There was a bold proposal from the U.S. and Canada to revise the then existing "Nutritional Requirements" and to establish a universal "Dietary Reference Intake" . At the symposium, there was a lively discussion on the significance and methods of the proposal. Six months later, Dr. Hosoya invited Dr. Vernon Young, who was the chairman of the "Review Committee on the U.S. Dietary Reference Intakes", to Japan, and a closed meeting was held with a small number of participants. Conventional nutrient requirements were recommended to prevent nutrient deficiencies, but the problem of over-nutrition and over-consumption of nutrients from supplements had emerged, and the necessity of establishing reference values that could be adapted to prevent both deficiencies and excesses was discussed. However, since it is difficult to determine appropriate reference values for individuals or a limited group of people, it was proposed that the reference values should be based on scientific evidence and estimated to reduce the risk of deficiency or excess. This is the origin of the current "Dietary Reference Intakes for Japanese".

At this conference, Dr. Young made a memorable statement. "There are many eminent nutritionists in the world, each an expert in his or her own nutrient, but none have been able to answer the basic question, 'What should we eat and how much? This is a major question for future nutritional research."

4.3   Inauguration of the "Committee to Study the Role of Dietitians in the 21st Century"

In Japan, as in Europe and the United States, the need for research and education centered on human nutrition has gradually spread, and people who share this philosophy have emerged, giving rise to new academic societies.

In 1980 (Showa 55), "The Japanese Society for Clinical Nutrition" was founded mainly by physicians, and the following year, in 1981 (Showa 56), "The Japanese Clinical Nutrition Association" was founded based on the collaboration between physicians and dietitians who actually take charge of clinical nutrition. On the other hand, in the surgical field, studies on nutritional supplementation associated with surgical intervention began, and in 1998 (Heisei 10), the "Japanese Society for Parenteral and Enteral Nutrition (JSPEN)" (At present, Japanese Society for Clinical Nutrition And Metabolism) was established. In 1998, the "Japan Society for Metabolism and Clinical Nutrition" was established based on internal medicine.

4.3.1   The Need for Clinical Dietitians

The doctors and dietitians who had received training in Europe and the United States strongly felt the need to establish a system of clinical dietitians stationed in hospital wards as in Western countries, and took every opportunity to gather information from Europe and the United States and to hold repeated training sessions to educate and enlighten the public in Japan. At conferences, we presented invited lectures by clinical dietitians from abroad We set up the "Nutrition Therapy (NT) Study Group", mainly in Tokyo and Osaka, and repeated study sessions at a pace of once a month. In the study group, we referred to the core curriculum of the American Society for Parenteral & Enteral Nutrition (ASPEN) for Nutrition Support Dietetics. This text was used as a reference (Photo 13). Based on this text, the first book on human nutrition written in Japanese, "Clinical Nutrition Management," was published. (Photos 4.1 and 4.2).

Research and training in nutrition and nutritional supplementation methods began to be carried out mainly by the Nutrition Therapy Study Group. In the course of these discussions, it was suggested that dietitians should be in charge of cooking, menu planning, and general nutritional guidance, and that the position of clinical dietitian to provide nutritional management and advice based on the assessment and evaluation of the nutritional status of patients.

From September 2 to 4, 1994 (Heisei 6), volunteers gathered at Akane-so, Ministry of Health and Welfare, under the title of "Study Group on the Clinical Dietitians System", and held discussions for 3 days and nights (Photo 15). The members who gathered were people who had trained in the US and Australia, members of the NT Study Group, those who were actively engaged in clinical nutrition activities, and also registered dietitians in the U.S. who had studied in the

Photo 4.1 Textbook for Nutrition Support Dietetics in the USA. (Gottschlich et al. 1993)

U.S. and were engaged in clinical nutrition work. Dr. Hosoya and the author took the lead in preparing a preliminary draft of the "Clinical Dietitians System" and published it in "Nutrition Japan, 37(12), extra issue" (Photo 4.3).

Dr. Hosoya requested the board of directors of The Japanese Society of Clinical Nutrition to establish "Clinical Dietitian" as a qualification system certified by the society. However, the discussion at the board meeting was confused, and in the end, it was adjourned because it was not possible to summarize the various opinions. This was because the board members did not understand the significance of creating the specialization of clinical dietitians and the nature of their work. The directors at that

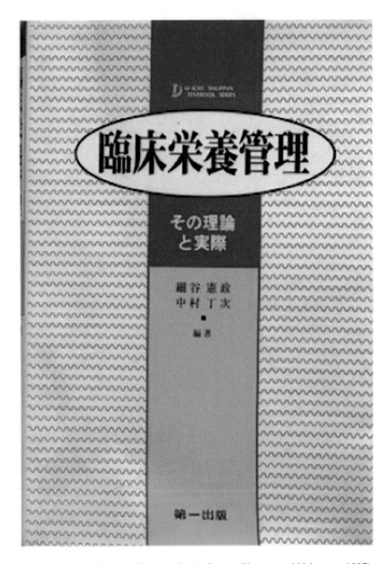

Photo 4.2 Textbook for Nutrition Therapy Study Group. (Hosoya and Nakamura 1997)

time were physicians and scholars who had academic interests in nutrition in the fields of physiology, biochemistry, cardiology, maternal and child health, pediatrics, metabolism, surgery, etc. It seems that they did not fully understand the significance of nutrition therapy and the training of professionals in clinical practice. Some people thought that it was not the job of academic societies to train professionals because nutrition research in Japan was mainly focused on agriculture and home economics. Dr. Yuichiro Goto, former president of Tokai University, who was the chairman of the study committee, Dr. Hosoya, and I were disappointed that our

Photo 4.3 Akane-So meeting where the clinical dietitian system was discussed (September 1994)

dream had ended in a mirage. However, now that I think about it, it is an undoubted fact that this kind of movement was the catalyst for the subsequent Law Reform 2000, and the members who participated in this movement became the leaders of human nutrition and clinical nutrition in Japan.

When discussing the clinical dietitian system, one of the topics that came up frequently was the role and duties in clinical practice. If we were to do the same thing as before, that would not be considered clinical work, and there would be no need to create a new qualification. Training that does not have specific duties and roles is at high risk of falling behind, as was the case with the "National Training Seminar on Pathological Nutrition" that we conducted previously. As a result of our discussions, we arrived at the concept of "nutritional management based on the evaluation and determination of the nutritional status of patients". The term nutritional management has been used in hospital food service, but this was the management of nutrients contained in the menu. The new clinical nutrition management was defined as nutritional management to improve the nutritional status of human beings, and its goal was to implement nutrition therapy, nutritional supplementation, and also nutritional education in order to improve the nutritional status of the subject. This was the exact embodiment of human nutrition.

This led us to the conclusion that the first thing we needed to learn in the future was "Nutritional Assessment", which is the evaluation and assessment of nutritional status. Around 1993 (Heisei 5), several volunteers from the Ministry of Health, Labor and Welfare, the National Institute of Nutrition, and St. Marianna University School of Medicine started a voluntary study group. The content of the study was to decipher Rosalind Gibson's thick "Principles of Nutritional Assessment". I

Photo 4.4 "Nutrition Japan", in which the clinical nutritionist system was published (vol. 37, no. 12, 1994)

remember reading the book with excitement, as it opened up a new world of nutrition that I had never encountered before. In fact, the members of this group became the key players behind the Law Reform 2000. Using Gibson's book as a reference, I published Japan's first review on "Nutritional Assessment" in "Nutrition and Dietary Life Information", the journal of the Nutrition Guidance Institute of the Japan Dietetic Association. I argued that the new role and duties of dietitians were to assess and judge nutritional status comprehensively, not only from the nutrient intake from the diet, but also from the body composition, clinical examination, and self-perceived symptoms.

4.3.2 Proposal of the Committee

Against this background, in 1997 (Heisei 9), the Ministry of Health, Labour and Welfare established the "Committee to Study the Role of Dietitians in the 21st Century" (Chairperson: Norimasa Hosoya). At this time, I remember Dr. Hosoya saying happily, "The Japanese Society of Clinical Nutrition has abandoned us, but the Ministry of Health, Labour and Welfare is going to create clinical dietitians." Representatives from a variety (Table 4.1) of fields participated in the committee, and after 1 year of extensive discussions, the following recommendations were made.

"Measures against lifestyle-related diseases have become a major health issue for the nation. In order to prevent the onset and progression of lifestyle-related diseases, it is important to improve dietary habits. And nutritional guidance requires a high

Table 4.1 List of the committee to study the role of dietitians in the twenty-first century

	Yoshiko Kagawa	President of Kagawa Nutrition University
	Mariko Kaneda	Director of Tama-Tachikawa Public Health Center and former Director of the Elderly Health Division of the Tokyo Metropolitan Government
	Noriko Kimoto	Critic
	Akihiko Koike	Former permanent member of the Japan Medical Association (1st to 7th)
	Shiro Goto	Professor Emeritus of the Tokyo University of Agriculture
	Hideya Sakurai	Permanent member of the Japan Medical Association (8th and onwards)
○	Hiroko Sho	Professor of the Open University of Japan, member of NHK management committee, former Vice Governor of Okinawa Prefecture
	Yutaka Seino	Professor of Kyoto University Faculty of Medicine
	Shigeyoshi Teramoto	Director of the Nagasaki Medical Central
	Yukihiro Nakabo	Professor of Kyoto Prefectural University
	Sumiko Nakamura	Food journalist representative and former chief producer of Nippon Television Network Corporation
	Teiji Nakamura	Director of Nutrition, St. Marianna University School of Medicine Yokohama Seibu Hospital
	Yoshitomo Fujisawa	Professor of Jissen Women's Junior College
	Makiko Fujiwara	Assistant city manager of Joetsu City
◎	Norimasa Hosoya	Professor Emeritus of the University of Tokyo
	Kazuki Matsumoto	President of Tokyo Shokuryo Dietitian Academy
	Masamizu Mizuma	Professor of Department of Physical Therapy, Showa University College of Medical Sciences
	Yasutoshi Muto	Professor of Sugiyama Jogakuen University

◎ Chairman, ○ Deputy chairman

level of expertise and skills based on nutritional assessment and evaluation. However, current registered dietitians are mainly involved in food service management, and few are involved in nutritional management for injured and sick patients based on nutritional evaluation and assessment. In Europe and the United States, registered dietitians are positioned as specialists in nutrition for people who suffer from chronic diseases and other illnesses, from prevention to treatment, so it is necessary to comprehensively review the role of registered dietitians in Japan."

In other words, before creating clinical dietitians, in order to rebuild the dysfunctional clinical nutrition system, registered dietitians should be educated and trained based on human nutrition, and their role should be to improve the nutritional status of humans, and their methods should include interpersonal work that introduces a management system. The study specifically concluded that it is necessary to introduce nutrition care management to improve the nutritional status of subjects with diverse and complex nutritional needs by assessing and evaluating nutritional status, planning and implementing appropriate diet, nutritional supplementation, and nutrition education, and monitoring and reassessing the results.

4.4   Innovations After the Law Reform 2000

As the study committees continued to discuss the issue, we wondered how best to frame the dietitian system, and furthermore, since the Dietitian Act was a parliamentary law, how to persuade parliamentarians to approve it as well. The Japan Dietetic Association was also involved in a number of discussions. Many parliamentarians and administrative officials helped in the revision of this law. In particular, I will never forget the presence of Takumi Nemoto, Deputy Minister of Health, Labour and Welfare (Minister of Health, Labour and Welfare from 2018). We consulted with him many times and he organized the requests from the Dietitians' Association and the issues related to the system, which were presented as the "Nemoto Memo" on July 16, 1999. After consulting with legal experts, he said that this would be a good idea.

"Nemoto Memo"
July 16, 1999

1. Since the introduction of dietitians in 1962
     Licensed nutritionist, registered dietitian
2. The role of the dietitian and the requirements for licensure have since become social realities appropriate for licensure

   ① Individual, interpersonal and professional nutritional guidance has replaced group meal guidance

(continued)

② The work of the dietitian will have an exclusive effect on the business
   Nutritional guidance fee by dietitian can be calculated from the
   medical service fee
③ Passing the national examination is now required to become a registered
   dietitian

3. Uncertainty about the scope of work is eliminated
4. Growing discrepancy between directions and reality, resulting in social
   confusion

In other words, the amendment aimed to clarify the duties of registered dietitians, to change the system from "registration" to "license", and to increase the effectiveness of individual and personal nutrition guidance. In order to do this, it was now essential to have education and training based on human nutrition and to pass a national examination. All of this was made possible by the "Law Reform 2000".

On July 21, 1999 (Heisei 11), the General Conference on the Reform of the Dietetic Act was held at the Hotel Okura in Tokyo, with delegations from each prefecture. The venue was filled with about 500 participants, and almost all the Diet members came to greet the participants and expressed their support for the revision of the Dietitians Act (Photo 4.5).

Photo 4.5 General Conference on the Reform of the Dietetic Act at Hotel Okura, Tokyo (July 21, 1999)

On March 15, 2000 (Heisei 12), The Dietitians Act was debated in the House of Representatives'Committee on Health and Welfare in the 147th Diet, following a report by the Committee to Study the Role of Dietitians in the twenty-first century. (Table 4.2). On April 7, the 147th ordinary session of the Diet approved a partial amendment to the Dietitians Act [effective 1 April 2002 (Heisei 14)].

The profession of dietitian was changed from a registration system to a licensing system, the qualifications for taking the examination were reviewed, and the new definition and duties of dietitians were clarified (Table 4.3). The conventional cooking, menu planning, and general nutritional guidance are [now] performed by dietitians, while nutritional management and guidance based on the evaluation and assessment of the nutritional status of the patient are performed by registered dietitians. The work at hospitals and other facilities centered on food service management has evolved into comprehensive clinical nutrition management, including not only food service management but also nutritional supplementation using catheters. The curriculum for training dietitians was also completely revised, and emphasis was placed on medical education in physiology, biochemistry, anatomy, pathology, and clinical nutrition.

Table 4.2 Proposed amendments to the Dietitians Act

In view of the importance of improving dietary habits in preventing the onset and progression of lifestyle-related diseases, the registered dietitian system should be reviewed

① A registered dietitian shall be positioned as a person who provides nutritional guidance, etc. necessary for the medical treatment of injured and sick persons. Nutrition guidance shall be provided under the guidance [direction?] of the attending physician

② The qualification of registered dietitians should be changed from a registration system to a licensing system

③ Reviewing the qualifications for registered dietitians and further upgrading their knowledge and skills as registered dietitians

Table 4.3 Definition of a registered dietitian

A registered dietitian is a person who is licensed by the Minister of Health, Labor and Welfare and works under the title of registered dietitian to provide nutritional guidance necessary for the medical treatment of injured or sick persons, nutritional guidance for the maintenance and promotion of health that requires a high level of specialized knowledge and skills in accordance with the physical condition and nutritional status of individuals, food service management that requires special consideration in accordance with the physical condition, nutritional status, and clients at facilities that continuously provide meals to a large number of specified persons, and guidance necessary for the improvement of nutrition at these facilities. Moreover, registered dietitian is a person who provides food service management that requires special consideration in accordance with the physical conditions, nutritional status, and conditions of users at facilities that continuously supply meals to a specified number of people, and providing guidance necessary for nutritional improvement of these facilities

4.4.1 *From the Nutrition Improvement Act to the Health Promotion Act*

In 2002 (Heisei 14), the "Nutrition Improvement Act" was amended to [into] the "Health Promotion Act". The Health Promotion Act focuses on measures to prevent lifestyle-related diseases, and nutrition issues are dealt with as part of comprehensive health promotion, including exercise, smoking cessation, and stress management. Under the Health Promotion Act, specified food service facilities must have a registered dietitian, and facilities other than specified food service facilities must also endeavor to have a registered dietitian. In these standards, the importance of assessment of nutritional status and of quality control of meals based on nutritional assessment of the subjects was described in accordance with the purpose of the revision of the Dietitian Act (Table 4.4).

4.4.2 *Changes in Dietetic Work*

The role of registered dietitians in medical and welfare care has changed dramatically since the "Amendment of the Dietitians Act 2000". Nutritional management in hospitals and welfare facilities has changed from the nutritional management of menu preparation to the nutritional management of injured and sick patients. In other words, the role of nutrition management in hospitals and welfare facilities has changed from the regulation of energy and nutrients contained in food to the management of nutrition to improve the nutritional status and health of people.

Table 4.4 Nutrition management standards in the enforcement regulations of the health promotion act

(Standards for Nutrition Management)
Article 9 The standards specified by an Ordinance of the Ministry of Health, Labour and Welfare referred to in Article 21, paragraph 3 of the Act shall be as follows
(i) The physical condition, nutritional status, lifestyle, etc. (hereinafter referred to as "physical conditions") of persons who receive meals (hereinafter referred to as users) through the said specified meal service facilities shall be periodically ascertained, and based on such information, efforts shall be made to provide meals that meet the appropriate calorific and nutrient content and to control the quality of such meals, as well as to evaluate such information
(ii) Efforts shall be made to prepare meal menus, taking into consideration the daily food intake, preferences of users, as well as their physical conditions
(iii) Information concerning nutrition shall be provided to users by posting the menu list and indicating the calorific value and main nutritional ingredients such as protein, fat, and salt
(iv) Preparation of menu lists and other necessary reference materials, etc. in an appropriate manner, and maintaining them at said facility
(v) Sanitation shall be managed in accordance with the provisions of the Food Sanitation Act (Act No. 223 of 1947) and other relevant laws and regulations

Conventionally, hospital food was categorized into special treatment food and general food. Special treatment food was prepared based on the doctor's dietary prescription for each patient, and general food was prepared by considering hospitalized patients as a group, calculating the weighted average recommended nutritional amount based on the characteristics of the group, preparing a menu that met the amount, and providing food using a group cooking method.

In addition, even for patients who were eligible for therapeutic diets, the nutritional reference amount was intended for the treatment of diseases, and the diets were provided without sufficient consideration for the nutritional status of the patients. Therefore, patients whose nutritional needs increased due to the stress of the disease became undernourished as a result of the therapeutic diet, and their taste buds changed as a result of the symptoms of the disease or the side effects of the [drop] drugs, and patients with reduced appetites left food uneaten, further worsening their nutritional status. In other words, the food service system in hospitals and welfare facilities at that time was not designed to improve the nutritional status of the injured, disabled, or individual patients. As a result, various types of malnutrition appeared, and this led to a decline in the quality of medical care and nursing care.

In order to solve these problems, "additional nutrition care management" by dietitians was approved for the first time in 2005 (Heisei 17) in long-term care insurance. This achievement was developed into the "additional fee for nutrition management" for medical care in the 2006 (Heisei 18) revision of medical fees. Although nutritional management is necessary for all inpatients, at the time, only a few dietitians had mastered the techniques of clinical nutritional management, and therefore reimbursement was granted only when it was implemented. In 2012 (Heisei 24), the "additional fee for nutrition management" was abolished and nutrition management was included in the calculation requirements of the basic hospitalization fee.

In other words, clinical nutrition management now became mandatory for all hospitalized patients.

4.4.3 Launch of the NST

In 2010 (Heisei 22), "Nutritional support team: NST" was approved for acute stage patients with injuries and illnesses for which nutritional management is difficult for dietitians alone, through multidisciplinary cooperation. The concept of team care was not limited just to nutritional management, but also extended to the treatment of bedsores, diabetes, kidney disease and cancer. In the area of health care, guidance for high-risk individuals in terms of preventing lifestyle-related diseases, in other words, measures against metabolic syndrome, was launched in 2008 (Heisei 20) as a specific health checkup and specific health guidance, in which registered dietitians participated together with doctors and public health nurses.

In just the 10 years after the Law Reform 2000, registered dietitians were placed in all fields of health care (primary prevention), medical care (secondary prevention),

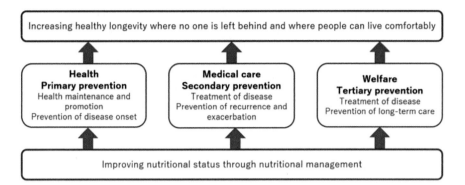

Increasing healthy longevity where no one is left behind and where people can live comfortably

Health
Primary prevention
Health maintenance and
promotion
Prevention of disease onset

Medical care
Secondary prevention
Treatment of disease
Prevention of recurrence and
exacerbation

Welfare
Tertiary prevention
Treatment of disease
Prevention of long-term care

Improving nutritional status through nutritional management

Fig. 4.1 The mission of the registered dietitian

and welfare (tertiary prevention) (Fig. 4.1). In a sense, a brand design for "Japan Nutrition" has been created. The Law Reform 2000, which required the development of nutrition policy based on human nutrition, was a major change in the education, work, and social evaluation of dietitians. It was also a major project that put our nutrition on a par with that of Western countries. Human nutrition has also influenced public nutrition, applied nutrition, and nutrition education since then.

4.5 Nutrition Management and Nutrition Care Process

The development of human nutrition has shown that the assessment of nutrition in the clinical field is not only a matter of developing a diet for an individual disease and assisting drug or surgical therapy. It has become clear that improving the nutritional status of the injured and sick can improve the therapeutic effect of drugs and surgical treatment, reduce the exacerbation of disease and the appearance of complications, decrease days of hospitalization, and reduce the increase in medical and nursing care costs. Traditionally, nutrition management in hospitals was limited to the management of hospital food, and the aim of this management was to ensure that the nutrients contained in the food were properly controlled, and that [drop] nutrition management was part of food service management. In other words, the nutritional status of each hospitalized patient was not assessed and judged before meals were served. Nutrition management was not designed to improve the nutritional status of patients.

4.5.1 Introduction of the NCP

Starting in 1990, there was a worldwide debate on the significance and methods of nutritional status assessment, determination, nutritional supplementation, and team care, and the question of the need for clinical nutrition management was raised. However, the methods differed among medical institutions and countries, and confusion continued. Under these circumstances, standardization of nutrition management was considered. In 1998, Polly Fitz, president of the American Academy of Nutrition and Dietetics (AND), established the Task Force on Nutrition Management in Health Services Research and began a full-scale study of nutrition management in 2001. In 2003, the AND formally decided to introduce the Nutrition Care Process (NCP) based on the results of the study and published the results in its journal.

The NCP is a "quality nutrition management system" for improving human nutrition and consists of the following:

1. Nutrition Assessment
2. Nutrition Diagnosis
3. Nutrition intervention
4. Nutrition Monitoring and Evaluation

In other words, the cycle begins with the evaluation and determination of the nutritional status of the subject, the formulation and implementation of an intervention plan, the monitoring and reevaluation of the results, and the continuation of further intervention. In Japan, this method was also introduced and developed as "nutritional care management" in the revision of long-term care insurance in 2005 (Heisei 17).

4.5.2   Proposals for International Standardization

AND proposed further international standardization of the NCP in 2005, and held the International Meeting on Standardized Language for Dietetics at AND headquarters in Chicago on August 23–24 of that year (Photo 4.6). The members of the meeting consisted of representatives from the United States, Canada, Israel, Australia, the United Kingdom, and Japan, and the present author participated in the meeting. At the conference, active discussions were held on the medical system, the situation of hospital food and nutrition management, implementation methods, education and training systems, as well as the nutritional status and nutritional problems of each country. Among these topics, a lot of time was spent on the introduction of "nutritional diagnosis".

The reason for this was that in many countries, there is a strong perception that "diagnosis" is an act that only doctors are allowed to perform because it involves responsibility for treatment methods, and there was a opinion in countries other than the United States that we should proceed with caution. I also talked about the current situation in Japan, and promised for the time being to publish a translation of the textbook used in the U.S. instead of introducing the NCP immediately. In the end, it was agreed that a symposium on NCP would be held at the ICD in Yokohama in 2008, and that the participating countries would work to educate and promote NCP in their respective countries, with the International Federation of Dietitians at the core.

**First International Meeting
in Chicago** August 23-24 2005

Attendee List

1) Australia
   Sandra Capra
2) Canada
   Marsha Sharp
3) Israel
   Naomi Trostler
4) Japan
   Teiji Nakamura
5) Dutch
   Jose Tiebie
6) Britishi
   Judith Catherwood
7) WHO
   Randa Jarudi Saadeh
8) NCHS
   David Berglund
9) SNOMED International
   Debra Konicek
10) America

Attendees

1 Australia Sandra Capra   **2 Canada** Marsha Sharp

3 Israel Naorria Tristler   **4 Japan** Teiji Nakamura

5 The Netherlands Jose Tiebie   **6 United Kingdom** Judith Catherwood

7 WHO Randa Jardi Saadeh   **8 NCHS** David Berglund

9 Systematized Nomenclature of Human Medicine (SNOMED)

   Debra Konicek

10 United States

Photo 4.6   Member of the international meeting on standardized language for dietetics

4.6   The Significance and Methods of Nutritional Diagnosis

The Nutrition Care Process (NCP) is a systems approach to providing quality nutrition care and is a standardized framework of how to manage nutrition. In implementation, it is created for each individual subject and not all patients/clients will receive the same nutrition and diet therapy. It is necessary to take into account the individual needs and characteristics of the patient/client, and to base the treatment on scientific evidence, and the key to this is the Nutrition Diagnosis.

4.6.1   Definition of a Nutritional Diagnosis

Nutrition diagnosis is the diagnosis of a subject's nutritional status based on the nutritional assessment and the identification of specific issues that should be resolved or improved by nutritional intervention. In other words, while the nutritional assessment evaluates each of the following items: food/nutrition history, biochemical data, medical tests, procedures, physical measurements, physical signs, and treatment history, the nutritional diagnosis is a comprehensive evaluation and judgment based on the individual evaluations of the nutritional assessment. For example, just as a physician comprehensively evaluates each patient's interview, physical signs, subjective symptoms, and clinical examination, and finally diagnoses the disease as "○○ disease", nutritional diagnosis aims to express the nutritional status in a single word using standardized criteria. Just as there are international diagnostic standards for diseases, it can be said that nutritional diagnosis has created international standards for determining nutritional status. In this way, interprofessional variation in nutritional management can be minimized, and the condition of the patient can be instantly understood by hearing the diagnosis, and nutritional issues that need to be resolved can be objectively determined.

It is important to note that a nutritional diagnosis is not the same as a medical diagnosis performed by a physician. In other words, a nutritional diagnosis is a diagnosis of a condition or phenomenon that is limited to the nutritional domain, and is based on the assumption that it can be improved by nutrition therapy intervention. For example, "energy/protein deficiency" or "beriberi" is a diagnosis of disease made by a physician, but a nutritional diagnosis is a diagnosis of nutritional status. When there is an insufficient intake of energy, protein, or vitamin B_1, and the nutritional status can be expected to improve by increasing the intake of the nutrient, the nutritional diagnosis is "energy/protein deficiency" or "vitamin B_1 deficiency".

AND has developed diagnostic criteria for 70 different types of nutrition problems, and the diagnostic criteria consist of the following three items:

① Intake: Excessive or inadequate intake of food or nutrients compared to actual or estimated requirements
② Clinical nutrition: nutrition problems related to pathological conditions and physical conditions
③ Behavior and environment: knowledge, attitudes, and beliefs of the subject, the environment surrounding the body, access to food, and food safety issues.

4.7 Writing Nutritional Diagnoses and Nutritional Interventions, Development of Monitoring

4.7.1 PES and Nutritional Diagnosis

Nutrition diagnosis is described by the standardized PES method: P (Problem or Nutrition Diagnosis Label) indicates the problem or nutrition diagnosis and what the patient or client needs in order to improve,;E (Etiology) indicates the causes and triggers that worsen the nutritional status,; and S (Signs/Symptoms) are the symptoms and signs of the subject and are the data in the nutrition assessment that are essential to the nutrition diagnosis (Table 4.6). The nutritional assessment is included in the S and the nutritional diagnosis is made by comprehensively evaluating these items. The nutritional diagnosis can be described in one sentence, "Based on the evidence of [S], [E] is the cause, the nutritional diagnosis is [P]". If expressed in this way, medical personnel can commonly understand the basis for the nutritional diagnosis and the factors that worsened the nutritional state, can recognize the contents of the most important nutritional disorder for nutritional management, and can implement high priority nutritional management.

For example, if a patient has lost weight due to a decrease in intake, a nutritional assessment will show that "the patient has a low eating rate and has lost 5 kg of weight over the past 4 weeks". However, this does not allow us to know "what caused the decrease in intake". In other words, an intervention plan cannot be established at this point. In the PES description, S is the nutritional assessment, E is the cause or trigger, and P is the nutritional diagnosis.

The nutritional diagnosis could be listed as "NI-2.1 Inadequate Oral Intake". The full description by the PES would be "due to an average 30% decrease in eating rate and 5 kg weight loss over 4 weeks (S), a decrease in appetite resulting from ill-fitting dentures and constipation (E), Inadequate oral intake (P)." (Tables 4.5 and 4.6)

4.7.2 Nutritional Interventions

The next step in the process, "nutritional intervention," is to develop a nutritional plan on how to modify and improve the diet and nutritional support to solve E. There

Table 4.5 Nutrition diagnosis items

NI (Nutrition Intake)	NI-1~5, subdivided into energy, water, nutrients, etc.
NC (Nutrition Clinical)	NC-1~3, subdivided into functional, biochemical and weight
NB (Nutrition Behavioral/ environmental)	NB-1~3, subdivided into Knowledge, Physical Activity and Food Safety and Access
NO (Nutrition Others)	NO-1 only, no nutritional problems at present

Table 4.6 Nutrition Diagnosis

Description of the nutritional diagnosis in the PES
(P)Problem, or Nutrition Diagnosis Label
Indication of problems and nutritional diagnosis
Details to be modified in the context of the patient's or client's nutritional status
(E)Etiology
Causes/related risk factors
→Nutrition interventions (planning and implementation)
(S)Sign/Symptoms
Symptoms and signs of the subject and data for nutritional diagnosis in the nutritional assessment
→Nutrition monitoring and evaluation
Writing : Based on the evidence of "S", "E" is the cause or related matter, nutritional diagnosis is "P"

Table 4.7 Notes on the development of nutritional intervention plans

1	Establish priorities for intervention
2	Refer to scientific evidence-based guidelines
3	Set expected intervention outcomes
4	Discuss with the subject and caregiver
5	Identify nutrition intervention plans and strategies
6	Identify the time and frequency of care required
7	Identify the tools you need

are two types of nutritional plans: a therapeutic plan and an educational plan, and it is necessary to develop an appropriate nutritional intervention according to the condition and needs of the subject (Table 4.7). The nutrition plan consists of four components: ① food and nutrition provision, ② nutrition education, ③ nutrition counseling, and ④ coordination of nutrition care.

4.7.3   Monitoring and Re-assessment

The final process, monitoring or reassessment, is to evaluate whether or not the Ss on which the nutrition diagnosis was based have improved. In other words, the symptoms/signs and laboratory tests used in the nutritional assessment are evaluated to determine the extent to which these have changed as a result of the nutritional intervention. In this case, it is important to quantify the subject's improved status. If the monitored items have improved, it can be determined that the nutritional treatment has been implemented as planned, and the causes and factors that worsened the nutritional status have improved. However, if they have not improved, the reasons why they have not improved should be re-examined, and the initial process should be returned to and re-assessed. In this case, if the nutritional assessment, which was the basis for the nutritional diagnosis, is improving, but the cause is not improving, then the treatment plan in the nutritional intervention is inappropriate and a change in the plan should be considered. In this way, the nutritional status is gradually improved by rotating the management cycle, which is a method of clinical nutrition management.

4.8   Dr. Hosoya's Last Message

In April 2016 (Heisei 28), 16 years after the Law Reform 2020, there was a workshop on the future of practical activities in clinical nutrition, organized by the Japanese Society for Clinical Nutrition. Prior to the workshop, a special lecture was given by Dr. Hosoya (Photo 4.7). He asked the participants if he could sit because of the burden on his legs due to his advanced age, and I operated the Power Point beside him. He said, "The research and practice of clinical nutrition in our country are lagging internationally, and we have not yet caught up. Those concerned must make greater efforts." Surprisingly, in the latter half of his talk, he suddenly stood up and leaned forward, and remained standing until the end of his talk. The participants

Photo 4.7   The last lecture by Dr. Norimasa Hosoya

were so overwhelmed by the power of his speech that they were all speechless, and the hall was enveloped in silence for a moment, with no one asking any questions or expressing any opinions.

Four months after this lecture, Dr. Hosoya suddenly passed away. Looking back, this lecture was the last message from Dr. Hosoya who kept shouting "Let's bring nutrition back to human beings", and if an extraordinary nutritionist named "Hosoya Norimasa" had not appeared in Japan in the latter half of the twentieth century, the Japanese nutrition world would have sunk like a leaky boat. During his lifetime, I heard criticism and complaints from him about the politics the government, academic societies, and individuals every week, and sometimes I was fed up with them.

Now that I think about it, I wonder if this is a sign that the reform was so difficult and, in fact, is still unfinished.

Unlike the time when we aimed to revise the law, today the social evaluation of nutrition is incomparably higher, and its academic reputation, the environment for research and education, as well as the role of the profession, have been improved. However, what we must not forget is that this situation was made possible by the many sleepless nights and efforts of many of our predecessors. If we do not know the history, we cannot see the problems of the present, and we cannot open the way to the future.

Bibliography

Documents of International Meeting on Standardized Language for Dietetics, August 23–24, 2005, American Dietetic Association, Chicago

Gibson RS (1990) Principles of nutritional assessment. Oxford University Press, Oxford

Gottschlich MM, Matarese LE, Shronts EP (1993) Nutition support dietetics Core curriculum second edition. ASPEN

Hosoya N (2010) Necessity of human nutrition – from the viewpoint of evaluation. J Jpn Soc Nutr Food 63(6):287–297

Hosoya N (2012) Clinical nutrition, introduction to clinical nutrition. In: Nakamura T (ed) Human nutrition necessary for team medicine. Dai-ichi Shuppan, pp 2–28

Hosoya N, Nakamura T (1997) Clinical nutrition management, its theory and practice. Dai-ichi Shuppan

Japan Dietetic Association supervision (2018) In: Kido Y, Nakamura T, Komatsu T (eds) Nutrition management process. Dai-ichi Shuppan

Japan Dietetic Association Supervisory Translation (2012) In: Kido Y, Nakamura T, Komatsu T (eds) Nutrition diagnosis, "Manual of nutrition care process terminology for international standardization". Dai-ichi Shuppan, pp 197–337

Lance K, Pritchtt E (2003) Nutrition care process and model: ADA adopts road map to quality care and outcomes ,management. J Am Dietetic Assoc 103(8):1061–1072

Nakamura T (1992) Nutritional Assessment. Nutr Dietary Inf 6(1):7–26

Nakamura T (2006) International standardization of nutritional management and introduction of nutritional diagnosis. Clin Nutr 11(1):89–91

Nakamura T et al (2008) Provide dietitians the key elements needed to implement evidence-based dietetics practice in their practice, Workshop, abstract book of 15th international congress of dietetics

Suzuki M, Hashimoto K (eds) (2009) Chapter 9: Dietitians, the dual structure of the nutrition system, the Japanese structure of professional training. Tamagawa University Press, pp 165–183

Chapter 5
Team Medicine and Multidisciplinary Education

Abstract Nutrition has taken a leading role in team medicine. The concept of the Nutrition Support Team (NST) is a specific example and a pioneer in team medicine. There was a fundamental challenge in promoting Interprofessional Work. It was the lack of Interprofessional Education. Kanagawa University has four departments – Nursing, Nutrition, Social Welfare, and Rehabilitation – and has been training professionals based on education and research through multidisciplinary cooperation since its establishment. In today's complex and diverse health, medical and welfare systems, problems cannot be solved by one profession alone, and students are taught that it is necessary to demonstrate their comprehensive abilities through interprofessional cooperation.

Nutrition is always necessary for the foundation of life and health and has a strong relationship with other fields. Furthermore, it is impossible to maintain a comfortable QOL (quality of life) with nutrient supplementation alone, so it is necessary to work with a dentist to improve chewing and swallowing in order to eat well. In addition, cooperation with medical doctors, nurses, physiotherapists, occupational therapists, and public health nurses is also necessary to improve diet therapy and eating habits. The improvement of appetite and taste, which is the gateway to dietary therapy, involves various and complex factors such as medical condition, medications, environment, mental state, and nursing care.

Keywords Team medicine · Skill Mix · Study Group on the Promotion of Team Medicine · Nutrition Support Team (NST) · Interprofessional Work (IPW) · Interprofessional Education (IPE) · Human service

5.1 Birth and Development of Team Care

Since the latter half of the twentieth century, "team medicine" has been a topic of discussion whenever the nature of future medical care is discussed. I was invited to participate in a symposium many times at medical and medical-related conferences. However, the content of the discussion was usually the introduction of each profession that would form part of a team, the necessity of a team, and the division of work

© The Author(s) 2022
T. Nakamura, *Japan Nutrition*, https://doi.org/10.1007/978-981-16-6316-1_5

roles. At that time, the significance and inevitability of forming a team, and how to form a rational team were not discussed. The reason for this was that the work, systems, and training in each profession were not yet mature, and each profession was occupied with establishing its own specialties. Each training course was also fostered the belief that that its own profession was essential to healthcare and that its own development would improve healthcare in Japan. In other words, "dream team medicine" was repeated for many years, but only in discussions among symposiasts from different professions.

5.1.1   The Birth of Team Medicine and Interprofessional Work in Europe and America

In Europe and America, team medicine was implemented early on. The reason for this was not the "dream team medicine" discussed in Japan, but a methodology born out of necessity in terms of risk management. In the 1990s, the number of deaths due to medical errors in the U.S. ranged from 44,000 to 98,000 per year, and the cost reached $17 billion to $29 billion per year. A Presidential Advisory Committee was set up to deal with the problem. The cause of medical errors was not inadequate professional knowledge or skills on the part of each professional, but "lack of inter-professional communication". Teamwork had to be improved in order to prevent medical errors from occurring.

In the UK, in 1998, the Bristol Children's Hospital operating theater confessed to an unusually high number of child deaths in surgery. The government set up a secret commission of inquiry and submitted a report in 2001. The committee concluded that the reason for the high number of deaths was not inferior skill on the part of the doctors involved in the operations, but a lack of communication and teamwork among the staff, as well as an absence of leadership.

Thus, a philosophy was born that many medical professions should cooperate and collaborate in order to cope with highly advanced medical technology and diversi-fied patient needs. Team medicine and team care are called interprofessional work (IPW), and in the 1990s, the need for and methods of IPW were discussed in OECD countries. In these discussions, the issue was approached not only as a way to prevent the aforementioned medical errors, but also as a way to solve the shortage of doctors. In other words, a review of tasks that can be performed by non-physicians within the scope of the Medical Practitioners Act and the promotion of the division of roles among physicians and other medical personnel were considered. This movement was called Skill Mix (multi-professional collaboration). Moreover, skill mix was not just a mere division of work roles, but also included consideration of the delegation of authority and responsibility within the medical team. The debate then evolved to include how staff with different qualifications and abilities should be mixed within the team, the delegation of authority and the creation of new job functions.

5.1.2 Promoting Team Medicine in Japan

In Japan, the Ministry of Health, Labour and Welfare (MHLW) launched the "Study Group on the Promotion of Team Medicine" in August 2009 (Heisei 21). All medical treatment had been performed "under the direction of a physician", but the committee discussed expanding the discretionary authority of other medical personnel to carry out such treatment, thereby reducing the workload of physicians and improving the quality of medical treatment. The report completed in March 2010 (Heisei 22). In the report, team medicine was defined as "a wide variety of medical staff engaged in medical care, based on their high level of expertise, sharing objectives and information, sharing tasks while collaborating and complementing each other, and providing medical care that precisely responds to patients' situations". There are three specific benefits of team medicine.

① improvement of medical care and quality of life, such as early detection of diseases, promotion of recovery, and prevention of serious diseases
② reduction of burden on medical personnel by improving efficiency of medical care
③ improvement of medical safety through standardization and organization of medical care.

In order to promote team medicine, it was decided that (a) improvement of the expertise of each medical staff member, (b) expansion of the role of each medical staff member], and (c) promotion of cooperation and complementarity among medical staff should be the basic principles.

In response to this study group, in April 2010, the Ministry of Health, Labour and Welfare (MHLW) issued a director-general's notice "On the promotion of team medicine through collaboration and cooperation of medical staff" to clarify the role of dietitians in the nutrition field (Table 5.1). According to this notice, the duties of dietitians "include determining or changing the content and form of general diets, proposing special treatment diets, judging the appropriate timing of nutritional guidance, and proposing the selection and change of enteral feeding agents, after receiving comprehensive guidance from doctors". The report states that dietitians should actively participate in team medicine.

Once again, team medicine is interprofessional work (IPW) in medical care, and in recent years, this philosophy has been expanded not only to medical care but also to the fields of health and welfare, and the significance and methods of IPW have come to be actively discussed. In 2008, the "Japan Association for Interprofessional Education" was established, led by Eimei Takahashi, former president of Niigata University of Health and Welfare, and discussions in a variety of fields were renewed.

In order to promote IPW, three perspectives are considered: ①communication, ② information sharing, and ③ team management. A number of conditions have been identified as being necessary for good IPW implementation. It is a question of creating a situation in which people are always encouraged to speak up within the

Table 5.1 Promotion of team medicine through collaboration and cooperation among medical staff

(3) Registered dietitian

In recent years, with the aging of patients and the increase in the prevalence of lifestyle-related diseases, the role that can be played by the registered dietitian in the medical field as a specialist in the evaluation and judgment of nutritional management and nutritional guidance for injured and sick patients has become significant from the perspective of improving and maintaining the nutritional status of patients, preventing a decline in immunity, and improving the effectiveness of treatment and QOL. The role of nutritional specialists in the medical field is significant.
    Since the following tasks can be carried out by dietitians under the current system, it is desirable to make active use of dietitians.
    ① The content and form of general meals (regular meals) shall be determined or changed under the comprehensive guidance of a physician
    ② Advice to the physician regarding the content and form of special treatment meals
    (This includes the proposal of changes to the content of meals, etc.)
    ③ Comprehensive guidance for physicians regarding nutritional guidance for patients (critical path) in determining the appropriate timing for implementation, and in implementing it
    ④ Proposal to the physician about the selection of change in the type of enteral feeding agent to be used when enteral feeding therapy is performed

Notification by the Director-General of the Medical Affairs Bureau, Ministry of Health, Labour and Welfare, dated April 30, 2010

Table 5.2 Conditions for good IPW

1	Different opinions are encouraged and individual interests and ideas can be openly expressed
2	The uniqueness of each profession is recognized
3	Members are aware of the limitations of their individual expertise and team
4	Members should always be willing to consider opinions from other professions and outside parties
5	Things should not be interpreted in a way that suits the individual or the team
6	Recognition and respect for the expertise of other professionals
7	The ethical and moral consequences of the team's decision are considered

team, and their interests and ideas are recognized, as well as the uniqueness of each professional. In addition, it is important to recognize one's own limitations, to take into account the opinions of other professionals and outsiders, and to recognize and respect other professionals (Table 5.2).

After the war, many medical professionals were trained, but cooperation with other professions was rarely discussed. In other words, we did not know what their philosophy was, how they were trained, or what sort of knowledge and skills they had. This ignorance of other professions led to an excessive "obsession" with one's own profession, which in turn led to a "stiffening" of the situation.

The marked differentiation of medical specialties has had the danger of losing sight of the wholeness of the patient and of interpreting things in a way that is convenient for one's own profession, "seeing the disease but not the sick person". However, as people's values and lifestyles diversify, the medical care that patients desire is becoming more and more sophisticated and diverse, and it is becoming essential for medical staff with a high level of expertise to collaborate and complement each other appropriately.

5.2 IPW and Nutrition

5.2.1 NST and Total Parenteral Nutrition

IPW has been actively discussed in nutrition, and nutrition has taken a leading role in team medicine. The concept of the Nutrition Support Team (NST) is a specific example.

In 1968, Dr. Stanley Dudrick, of Harvard Medical School, developed a catheter-based method of total parenteral nutrition and established NST, a multidisciplinary team dedicated to the implementation and dissemination of this method of feeding. At the time, such an innovative method of nutritional support could not be implemented by doctors alone, but required the participation of relevant registered dietitians, nurses and pharmacists. In 1973, the first NST was officially established at Boston City Hospital in the United States. Around this time, Dr. George Blackburn's nutritional assessment was systematized, and in 1975, the American Society for Parenteral and Enteral Nutrition (ASPEN) was established with physicians, registered dietitians, nurses, pharmacists, and others as members. In other words, it can be said that the field of nutrition was a pioneer in team medicine for the practice and operation of innovative technologies for nutritional supplementation.

In the U.S., around 1990, NSTs were established in acute care hospitals to provide nutritional assessment, determination of nutritional support, and management by a team. For example, when NSTs were created, requests for total parenteral nutrition would come to them, but actually only half of them were implemented. This was because detailed nutritional assessment had reduced the excessive use of central venous nutrition. In addition, the use of inappropriate nutritional materials, catheter sepsis, and abnormalities in blood glucose and electrolytes were significantly reduced (Table 5.3).

5.2.2 Launch of JSPEN

In Japan, the Society for Complete Venous Nutrition was established in 1970 (Showa 45), the Japan Society for Venous and Enteral Nutrition in 1985 (Showa 60), and the Japanese Society for Parenteral and Enteral Nutrition (JSPEN)* in 1998 (Heisei 10), the latter led by former Vice President Shohei Kogoshi of Kochi University. In Japan, the establishment of an academic society to serve as a governing body for NST, as is the case in Europe and the United States, was delayed for more than 25 years. One of the reasons for this is that Japanese medical care as a whole had little interest in nutrition, and the reform in the training of dietitians who

Table 5.3 Qualitative changes in nutrition management with the creation of NST

Items	1990		After NST 1992~1993	
	n	%	n	%
Request for a TPN from a doctor in charge			208	
Patients who have been treated with TPN	77		122	59.0
Inappropriate nutritional supplements	19	24.7	1	0.5[*]
Catheter sepsis	8	10.0	7	5.7
Hyper -/hypoglycemia	19	24.7	6	4.9[*]
Hyper -/hypokalemia	3	3.9	0	0
Hyper -/hyponatremia	15	19.5	0	0[*]
Hyper -/hypophosphatemia	9	11.7	0	0[*]
Hyper -/hypomagnesaemia	5	6.5	0	0[*]

Data: Flsher and Opper (1996)
[*]<0.001

must play a central role had been delayed. In other words, it took a long time for dietitians to be freed from food service work.

*The Japanese Society for Parenteral and Enteral Nutrition changed its name to the Japanese Society for Clinical Nutrition and Metabolism in January 2020.

5.3   Interprofessional Education and the Challenge of Kanagawa University of Human Services

5.3.1   The Importance of Interprofessional Education

There was a fundamental challenge in promoting IPW in our country, which was the lack of Interprofessional Education (IPE) for this purpose. In 2010, WHO published a "Framework for action on interprofessional education and collaborative practice" and recommended interprofessional education to the world-wide (Fig. 5.1). According to this framework, it is necessary to improve the general ability of professionals and the ability of professionals to cooperate one another, as well as to improve the ability of each professional as before.

Especially in Japan, where the aging of society is advancing, inter-professional cooperation has become an essential issue as problems such as caring for elderly people who need nursing care, transition to community and home care, and controlling medical and nursing care costs arise, but education and research for this purpose are remarkably lagging behind the need . In recent years, discussions of IPE have become more active, and several competency models have been presented, but they generally consist of four domains (Fig. 5.2).

Gathering of professionals does not make for Interprofessional Work (IPW)
Interprofessional Education (IPE) is necessary

Interprofessional is different from Multi-professional in
that it refers to the interaction between professionals

Fig. 5.1 Encouraging interprofessional education

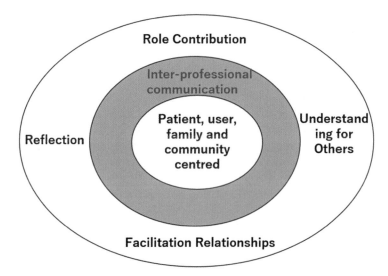

Fig. 5.2 Inter-professional competency model as a collaborative skill

Its contents can be explained as follows.

1. Fulfilling the role of the job: Role Contribution
   Each profession understands one another's role and can fulfill their role as a
   profession while utilizing one another's knowledge and skills.
2. Working on the relationship: Relationship Facilitation

Able to support and coordinate the development, maintenance and growth of relationships with multiple professions. Each profession is also able to respond appropriately to inter-professional conflicts that sometimes arise.

3. Reflection on one's own occupation: Reflection

Able to Reflect on the thoughts, actions, feelings, and values of one's own profession, and gain a deeper understanding of the experience of collaborative work with multiple professions, and apply this understanding to collaborative work.

4. Understanding of Others

Able to Understand the thoughts, actions, feelings, and values of other professions and apply them to collaborative work.

5.3.2   Education of Kanagawa University of Human Services

Kanagawa University has four departments – Nursing, Nutrition, Social Welfare, and Rehabilitation – and has been training professionals based on education and research through multidisciplinary cooperation since its establishment. In today's complex and diverse health, medical and welfare systems, problems cannot be solved by one profession alone, and students are taught that it is necessary to demonstrate their comprehensive abilities through inter-professional cooperation. The significance and methods of collaboration are taught in lectures, exercises, and practical training, and special research funds are allocated for research on multidisciplinary collaboration, in an attempt to change from the specialization seen in traditional universities. The curriculum is designed to provide education in each of the professions, and in the first year there is a symbolic subject of "Human Services", which includes a classroom lecture on the philosophy and necessity of human services from each department. In addition, the theory of health and medical welfare is taught in the first and second year, the theory of regional health and medical welfare is taken in the third year, and human service II is taken in the fourth year. (Table 5.4).

In "Health and Medical Welfare", students learn about the basic concepts and activities of each profession regarding the systems and activities that support health and medical welfare, and learn about the significance and necessity of collaboration, especially with the user at the center. In addition, students learn about the concepts, history, subjects, and fields of nursing, nutrition, social welfare, and rehabilitation studies, understand the current status and issues of each profession, and learn about the nature of cooperation. Students will then visit hospitals and social welfare facilities to learn about the actual practice of healthcare and welfare and about their users. In "Human Services II", students conduct mock case conferences using examples in the second semester of their fourth and final year, based on the expertise they have learned so far at the university. In other words, the objective is to acquire the ability to provide inter-personal support and assistance from the perspective of

Table 5.4 Kanagawa University of Human Services curriculum on IPE

First year	Second year	Third year	Fourth year
Symbolic subject			
Human service I			Human service II
General human education courses			
Relationships and communication human rights, gender and more			
Collaborative practical education courses			
Health and medical welfare Theory I	Health and medical welfare Theory II Counselling theory	Regional health and medical welfare Theory Welfare Collaboration Theory	Human service Comprehensive exercises
Specialized creative education courses			
Various specialist subjects Graduation research			

"Human Service" in practice after graduation. Specifically, the following methods are used.

① Cases and examples are presented for each of the mixed groups of students from the four departments.
② In each group, students evaluate the subject from their own professional standpoint and make a comprehensive assessment.
③ For each group, students work together to develop an individual support plan.

Steps ① ~ ③ are performed in an exercise format.

In addition, when the need for preparation or review arises, instructions will be given on study contents and methods as needed.

④ General comments from department chairs, faculty, and deans

As mentioned above, our university has been experimenting with collaborative education throughout the university. The results are unclear at present, but on the occasion of the 10th anniversary of our university, we conducted a questionnaire survey of our graduates. As a result, "Awareness of inter-professional collaboration" topped the list of "The things that students learned at university", and "Ability to be aware of inter-professional collaboration" was selected as a "Useful ability" (Table 5.5).

In recent years, many medical and welfare professionals, as well as educators, have presented the form of team medicine and care. There is no problem if such medical treatment and care become reality, but I have heard graduates of our university say, "In the field, it is not yet a team medical treatment" or "There is a gap between the ideal of education and the field". Recently, I have been thinking that the key word for the success of IPW and IPE is to have a "margin" as a "potential" to

Table 5.5 Results of a survey of Kanagawa University of Human Services graduates (Sixth Graduating Class: 2011)

1. The things that students learned at university		
1.	Awareness of inter-professional collaboration	72.7%
2.	Academic specialization	63.6
3.	Communication skills	43.2
4.	Human Services Philosophy	34.1
5.	The ability to contribute to teamwork	29.5
6.	Problem finding and solving skills	27.3
2. Useful abilities		
1.	Ability to be aware of inter-professional collaboration	61.9%
2.	Professional knowledge and skills	59.5
3.	The ability to contribute to teamwork	40.5
4.	Communication skills	35.7
5.	Problem finding and solving skills	23.8
6.	The ability to implement human services	21.4

work together with other fields. If you don't have a "margin", there is a risk that the overlapping area will be interpreted as an intrusion or invasion and the work in the boundary area will develop into a dispute. Rather than reaching out to other areas and expanding our own areas, we should aim to create margins where we can work together and improve the quality of our respective areas. To this end, we believe that the most important thing for the success of collaborative work is to respect and honor other professions, rather than ignoring or disrespecting other domains. Therefore, interprofessional work is meaningless unless the new results produced by the team are appreciated, and at the same time, the expertise of each team member advances. This means that it is not possible for dietitians alone to improve nutritional status.

Nutrition is always necessary for the foundation of life and health, and has a strong relationship with other fields. For example, it is necessary to cooperate with doctors, nurses, and pharmacists for various nutritional supplements. Furthermore, it is impossible to maintain a comfortable QOL (quality of life) with nutrient supplementation alone, so it is necessary to work with a dentist to improve chewing and swallowing in order to eat well. In addition, cooperation with physiotherapists, occupational therapists, and public health nurses is also necessary to improve diet therapy and eating habits. The improvement of appetite and taste, which is the gateway to dietary therapy, involves various and complex factors such as medical condition, medications, environment, mental state, and nursing care.

In the twentieth century, each profession developed by improving their professional education. However, in the twenty-first century, as health care, medical care, and nursing care rapidly become more sophisticated, problems are emerging that cannot be solved no matter how much effort is made by each profession. Under these circumstances, if we receive support from other professions and collaborate with other professions, we will be able to overcome the insurmountable barriers.

As is apparent, professional education for each of these professions is in a period of great change. In 2012, the Japanese Association of Nutritional Science Education was founded by Dr. Heizo Tanaka, former director of the National Institute of Nutrition, and the present author is currently the president of the society. It is my sincere hope that many people will join this society and study the future of nutrition and dietitian education.

Bibliography

Fisher GG, Opper FH (1996) An interdisciplinary nutrition support team improves quality of care in a teaching hospital. J Am Dietetic Assoc 96(2):176–178

Hosoya N, Nakamura T, Adachi K (2001) Dietary supplements, 'Health and nutritional foods' and nutritional management. Team Medical Care

Ministry of Health, Labour and Welfare (2010) Report of the Study Group on the Promotion of Team Medicine

Nakamura T (2012) Chapter 1, 3: Team medicine. In: Approaches to human nutrition necessary for team medicine. Dai-ichi Shuppan, pp 29–31

Nehme AE (1980) Nutritional support of the hospitalized patients – the team concept. JAMA 243:1906–1908

Otsuka M (2012) Chapter 4: Professional collaboration practice. In: Care and IPW (Inter-professional work) to support eating. Kenpaku-sha, pp 27–32

Toyama K, Nakamura T (eds) (2012) Chapter 5: Nutritional support team, Approaches to human nutrition necessary for team medicine. Dai-ichi Shuppan, pp 296–302

Chapter 6
Safe and Appetizing Patient Meals

Abstract Hospital malnutrition is caused by loss of appetite, taste, digestion, absorption of nutrients, and metabolic changes. Hospital Malnutrition decreases the QOL of patients, reduces the therapeutic effect of drugs and surgery, increases the length of hospital stay, and ultimately increases the cost of medical care. It is the necessary to reform hospital diets into delicious and comfortable diets and building a nutrition management system to solve the hospital malnutrition.

We created "patient cafeterias" where this room was decorated in orange, a color that stimulates appetite, background music was played. The temperature in meals were served hot or cold. Controlling the temperature of meals was essential not only to provide a tasty meal, but also to prevent food poisoning.

Operations such as clerical work and preparation in the kitchen that do not involve direct contact with patients were thoroughly IT, At or robot, while services on the front that were visible to people were provided with a human touch, and this was described as "softening on in front and hardening in back".

Keywords Medical Service Act · Complete Meal System · Standard Meal System · Inpatient Meal Treatment System · Temperature control · Patient cafeteria · Selection menu · Asian Conference of Nutrition · Asian Congress of Dietetics

6.1 History of Hospital Meals

Since the Meiji Restoration, the modernization of medicine and medical care has also had an impact on hospital food. Under the influence of Anglo-American and German medicine, the concept of patient meals based on modern nutritional science gradually came to be discussed. In 1888 (Meiji 21), Chiyokichi Hirano of Juntendo Hospital was the first to introduce Western-style meals for the sick, and published "A New theory of diet therapy" to include Japanese meals. Yukichi Fukuzawa founded Keio University Hospital with Dr. Shibasaburo Kitazato and opened the "Research Institute of "Diet Therapy "in 1926 (Taisho 15). They began a full-scale study of meals for the sick and formulated the concept of therapeutic meals, and the results of this study were developed into meals for hospitalized patients.

© The Author(s) 2022

T. Nakamura, *Japan Nutrition*, https://doi.org/10.1007/978-981-16-6316-1_6

However, the institutionalization of meals for hospitalized patients was largely due to the guidance of the United States after the war. In 1947 (Showa 22), the GHQ surveyed hospitals and pointed out to the government the need for improvement of medical care in Japan, which led to the enactment of the "Medical Service Act" the following year. As a result, medical institutions were established so that all citizens could receive medical care based on modern medicine, and hospital meals and hospital dietitians were included as part of this system.

6.1.1 From a Complete Meal System to the Standard Meal System

In 1950 (Showa 25), the "Complete Meal System" was formulated with the aim of ensuring that hospitalized patients would be able to obtain an adequate amount of nutrition from hospital meals alone, without having to bring supplementary food from home. In those days, when a patient was admitted to a hospital, it was common for him or her to be admitted with a pot and bedding (mattress, blanket, etc.). It was also common for hospitalized patients and their families to cook in a corner of the hospital room or in the corridor, and to bring in meals from home, without any hygiene or nutritional control. The complete meal system ensured that the patients would receive the necessary amount of nutrition per day from hospital meals alone. It would be unthinkable today, but all hospitalized patients were uniformly provided with 2400 kcal of food. The purpose of this policy was the strong desire of those involved in nutrition to improve the nutritional status of hospitalized patients, who were sickly and weak, by securing food for them on a priority basis during the severe food shortage after the war, when many people were dying of starvation and malnutrition. In a manner of speaking, all hospitalized patients were in a state of nutritional deficiency, and the meals were designed to solve this problem.

Later, as society stabilized and food supplies became dependable again, hospital food changed from securing quantity to improving quality, and in 1958 (Showa 33), the "Complete Meal System" was changed to the "Standard Meal System". Hospital meals were allowed to be added to medical service fees if they satisfied certain criteria set by the government, and hospital meals became an integral part of medical care as well as a qualitative improvement. The qualitative improvement of hospital meals at that time involved reducing the overemphasis on staple foods, such as a large amount of rice. Specifically, it aimed to improve the quantity and quality of side dishes, increase the intake of animal protein, and increase the intake of protein, fat, and even vitamins and minerals. Therefore, in the qualitative assessment, the most important factor was the "animal protein ratio" from meat, fish, eggs and dairy products to the total protein intake provided. The higher the proportion of animal products, the higher the quantity and quality of protein, and the higher the intake of vitamins and minerals. However, raising the "animal protein ratio" increased the cost of foodstuffs, so there was a battle between hospital managers who wanted to lower the cost of meals and nutritionists who wanted to ensure quality. However, there was a challenge to this index. Vegetarian hospitals, which had adopted vegetarian meals for religious reasons, were unable to raise this ratio and could not meet the standard for standardized feeding even though they provided finely-tuned meals using alternative foods such as soybeans.

6.1.2   Report on the Calorific Requirements of Hospital Meals for General Patients

In 1973 (Showa 48), the National Dietetic Council abolished the 2400 kcal nutrient amount that had been required up to that time, and changed the content so that meals could be provided that were closer to the appropriate amount for each individual patient. In other words, the "Caloric Requirements for Patients (15 years and older) on General Meals in Hospitals" was issued. The calorific value for patients provided with general meals in hospitals was calculated by adding a correction factor of 0.6 to the life activity index of 'nutritional requirements' for healthy people by sex and age. The intensity of activities of daily living of hospitalized patients was estimated to be 60% of that of healthy people leading their ordinary lives. Originally, there was a principle that meals for the sick should be determined on an individual basis, considering not only the sex and age of the patient, but also the patient's individual activity level, nutritional status, and the effects of his or her illness. However, in reality, hospitalized patients were managed as a group, similar to the practice for school lunches or industrial lunches, and individual attention was not possible.

6.1.3   Start of the Inpatient Meal Treatment System

In 1994 (Heisei 6), the "Standard Meal System", which was intended to ensure the quality of hospital meals, was abolished because its purpose had been achieved, and a new "Inpatient Meal Treatment System" was started, which included a partial flat-rate co-payment for meals. At that time, the importance of determining the amount of nutrition for each patient was pointed out, taking into consideration the patient's individual medical condition and nutritional status. The new system aimed to provide meals for sick people that corresponded to the individual needs of each patient, but unfortunately, the specific method for doing so was not provided at this time either. In other words, it was not clear how to respond to each individual patient, and it was necessary to wait until the revision of the law in 2000 (Heisei 12) to develop hospital food service into a part of clinical nutrition management as we know it today.

6.2 Hospital Meals Are Bad

There was a difficult problem that had to be solved before hospital food could be nutritionally managed in accordance with the needs of individual patients. This was to deal with complaints that hospital meals were bad. Malnutrition was observed in many hospitalized patients, and even when it was pointed out that the solution to this problem was important in terms of medical care, the answer from many people was that it was not a problem in terms of nutritional management, but that it was because the hospital meals were bad and patients could not eat them. As the malnutrition of the postwar period was resolved and people became more affluent, hospitalized patients began to chant, "Hospital meals are bad, bad, bad," and there was a strong demand for hospital food to be tasty.

Why was it claimed that hospital meals were 'bad', yet this problem had been neglected?'

In general, if a restaurant serves bad meals, customers will stop coming and the restaurant just 'will eventually close.' Therefore, cooks think of taste first and devise ways to improve it, and managers make efforts to improve service by scouting cooks who can prepare delicious food. Hospital meals were positioned as part of the treatment from the time of the Medical Service Act set up after the war, and when the Standard Meal was set up in 1958 (Showa 33), hospital meals were included in the social insurance medical fee in 1961 (Showa 36) an additional charge for special treatment meals was approved. Special meals subject to supplementation were defined as "therapeutic meals with nutritional amounts and contents corresponding to the patient's age, medical condition, etc., which are provided on the basis of a meal plan issued by a physician as a direct means of disease treatment "(Summary of social insurance medical fee for 1994 (Heisei 6)).

6.2.1   The Problem of the Producer

In other words, hospital meals were incorporated into the medical system, the significance of being part of the treatment was emphasized, and hospital meals were considered to be the same as medicine. If they were the same as medicine, there is a saying that "good medicine is bitter in the mouth," and it was recognized that hospital meals do not have to be tasty but should be eaten obediently for the sake of treatment, and dietitians and medical personnel neglected efforts to make meals tasty. Moreover, unlike customers in restaurants, hospital patients could not run away.

However, the affluent society that came with high economic growth created a gourmet orientation, and hospital meals were required to be tasty. At that time, the necessity of patient service was emphasized throughout the medical care system. Naturally, there was a demand that hospital meals be made appetizing, and mass media carried articles every day saying that hospital meals should be tasty. In those days, hospital meals were used as a metaphor: "This restaurant is as bad as hospital meals," so that people could understand how bad the restaurant was. At the time, former Prime Minister Kakuei Tanaka, who was arrested in the Lockheed affair and spent time in prison before being hospitalized for a stroke, commented that hospital meals were worse than prison food. Based on this social background, both the government and hospital officials started full-scale efforts to make hospital meal tasty.

What could we do to make the meals more delicious at St. Marianna University Hospital, where I participated? In the process of implementing the improvement measures, I suddenly thought of a question: "In the first place, are sick people in a condition where they can perceive the taste of food?" When a person is healthy, he or she has an appetite, can perceive the taste of food, and can eat as much as he or she wants. However, when the person becomes ill, these capacities are reduced and the patient is not able to eat well and the sensation caused by food is limited both in quantity and quality.

6.2.2 The Problem of the Eater

We investigated the sensitivity of taste in patients with liver disease. Different concentrations of sweet, salty, sour, and bitter liquids were dabbed on filter paper and placed on the tongue to determine the sensitivity to each taste. In the acute phase of the disease, all patients with liver disease showed a decrease in taste sensitivity. However, as the patients' condition improved, their taste sensitivity improved and they were able to eat with appetite. When a patient is first admitted to a hospital and feels that "hospital meals are bad," there is often a problem on the part of the eater.

6.3 Warm Is Warm, Cold Is Cold

Although hospitalized patients have a problem of decreased gustatory function in sensing taste, we thought that we should try to make hospital food more appetizing.

6.3.1 Improving Where Patients Eat

The first question we considered was whether the bed was an appropriate place for eating. In general, hospital rooms have a peculiar odor caused by disinfectant and body odor, and if there is a portable toilet under the bed, it is like eating in a toilet. No matter what kind of high quality food is served in a toilet, it will not taste good. In general, patients eat in bed, but once the acute symptoms have passed and a patient's condition has stabilized, I thought that the patient should eat in a place appropriate for eating.

In St. Marianna University Hospital, on each floor there was a pantry where meals were prepared. This space was converted to create the first "patient cafeterias" in Japan. The room was decorated in orange, a color that stimulates appetite, and background music was played. We also controlled the temperature to ensure that meals were served hot. The patients were given a choice between the bed and the cafeteria. By improving the food environment in this way, the number of leftover meals decreased and the amount of food consumed increased. Later, this method was approved as an "additional fee for patient cafeteria" in the medical treatment fee because it was useful for recovery from illness, and this method spread nationwide.

In 1987 (Showa 62), "St. Marianna University Yokohama City Seibu Hospital" was established in Yokohama. At the opening of the new hospital, we took on the challenge of creating a model for hospital meals in Japan.

In the new hospital, I was able to create the hospital meal service that I wanted because I started from a completely blank slate in terms of hardware such as the kitchen, office, and patient cafeteria, as well as of software such as the meal service system and computerization. Patient cafeterias were set up on all floors. In order to provide hot meals, heat-insulated wagons were introduced throughout. One wagon costing 2.5 million yen was purchased for each ward (Photo 6.1). At that time, the hospital director said to me, "Nakamura bought 15 expensive Japanese cars" It cost a lot of money to make the food tasty, but in the end the director of the clinic was pleased to hear the high evaluations from his patients.

In general, in Europe and the United States, expensive investments are made in the temperature control of meals, but in Japan, hospital management was not enthusiastic about the temperature of meals. This is because, in Japan, the provision of hot meals was carried out to suit patients who wanted to eat their food hot but in Europe and the United States, meal temperature control was treated as a risk management issue to prevent food poisoning.

This is because it is important to prevent food poisoning by keeping staple foods, main dishes and soups warm and side dishes, pickles and fruit cold, and avoiding leaving them at room temperature where bacteria and viruses can multiply. Temperature control is necessary even during cooking operations in the kitchen, and food poisoning will not occur if food is kept at a high or low temperature where bacteria do not multiply for the 1 or 2 h from immediately after the end of cooking until the food is put in the patient's mouth. In order to prevent the food from being left at room temperature, a heater and a refrigerator are placed along the tray line where the serving work is carried out, and a hot and cold delivery vehicle is used to carry the food to the wards so that the temperature can be controlled without interruption. In fact, I have managed hospital meal service for more than 30 years and have never had a single outbreak of food poisoning.

Controlling the temperature of meals is essential not only to provide a tasty meal, but also to prevent food poisoning. If food poisoning occurs in a hospital, the kitchen will be closed for at least a week, and the name of the hospital director will be

All paper work is computerized (above)

The menu is prepared on the train line (bottom left)

Hot and cold delivery vehicles (bottom right) deliver to the wards

Photo 6.1 Temperature control with hot and cold delivery vehicles. (All paper work is computerized (above); The menu is prepared on the train line (bottom left); Hot and cold delivery vehicles
(bottom right) deliver to the wards)

reported in the newspaper. Temperature control for hospital meals is an inexpensive investment in security from a crisis management perspective.

6.4   Softening in Front and Hardening in Back for a Pleasant Meal

In Europe and the United States, there was a system where people could choose their own menu for hospital meals. Why couldn't we do that in our country? I thought that since airplanes offer in-flight meals where you can choose from two kinds of warm food even in a small space at an altitude of thousands of meters, there is nothing that cannot be done in hospitals on the ground. In Western hospitals, when you are admitted to the hospital, you are given a menu list and can choose all the dishes, which is called the restaurant menu system. I first considered introducing this method. However, I realized that it would be too complicated and impossible in Japanese hospitals where patients are hospitalized for more than a month, although it is possible in Western hospitals where hospitalization usually lasts 2 or 3 days. This is because the repetition of one or two menu lists or a weekly menu limits the number of dishes that can be served and patients become bored. In addition, it was nearly impossible to keep the nutritional content of freely selected menus constant.

6.4.1   Introduction of a Selection Menu

What struck me was the selection menu system for in-flight meals on airplanes, where [maybe add 'only'] the main dish is selected. Even in daily meals, people generally choose what they want to eat, whether they want meat or fish, Western or Japanese food, whether they want rich or light food, etc., from among two types of main dishes. If we were to create two sets of menus with a selection of main dishes, including therapeutic meals, we thought this would be feasible if we used computers, which were becoming common at the time. The more we improved patient services such as offering selective menus, the more complicated the work would become and the more information would be required. However, we thought that we could solve the seemingly contradictory task of improving service and streamlining by actively using computers. Most of the clerical work involved in providing meals, such as the management of meal types, number of meals, and ingredients, the issuing of meal tags, the calculation of orders to vendors, and the preparation of order sheets, was all computerized In other words, the work behind the scenes of providing meals was thoroughly streamlined, and the philosophy of the business was to provide as much detailed human service as possible in direct contact with patients, such as calculating the appropriate amount of nutrition for each patient, collecting information on preferences, complaints and requests from patients, intake capacity, and providing advice. Operations such as clerical work and preparation in the kitchen that do not involve direct contact with patients were thoroughly computerized or robotized, while services on the front that are visible to people were provided with a human touch, and this is described as "softening on in front and hardening in back".

6.4.2   Computerization of the Hospital Meal System

At that time, in order to build a hospital meal system, we aimed to computerize as much information as possible and robotize kitchen operations. We already had an expert draw up an illustration of an unmanned kitchen (Fig. 6.1). This did not turn out as illustrated, but immediately after the opening of the hospital, visitors poured in daily.

In Japan, 'heart' is used to describe food that is prepared with care and concern for the person who will consume it. Therefore, the Japanese consider a meal that is reasonably prepared by humans to be a poor meal that is not heartfelt. In the first place, heart does not exist in a dish, but the person who eats it tastes it and feels the heart. Therefore, a dish should be made rationally so that the person who eats it can feel the heart, and a first-rate cook does not put his or her heart into each dish, but rather has mastered a technique that allows the person to feel the heart. It is more effective in terms of hygiene control if there are as few people as possible in the

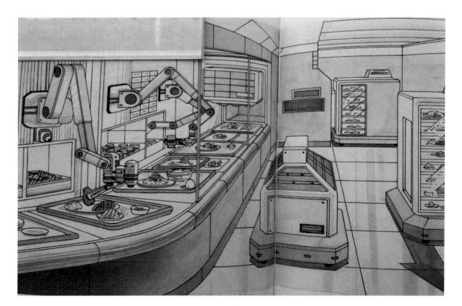

Fig. 6.1 Unmanned kitchen Created 30 years ago based on "softening in front and hardening in back"

"Documents: The future of hospital nutrition departments, Visual Dictionary of Nutritional Sciences, Salvio, Volume 2: Body and Nutrition, Dailec, pp. 162–163, 1988"

kitchen, because there is less chance of contamination by E. coli. At the hospital, I never said "cook with all your heart", but in the questionnaires I regularly received from patients, they always replied, "Thank you for your heartfelt cooking." At present, there is a lot of discussion about the professions that will decline and disappear with the advent of AI, but this kind of experience already took place in the era of the introduction of computers.

Referring to our challenge, in 1992 (Heisei 4), the government approved the "additional fee for specially managed meals" in the medical fee schedule for meals served at appropriate temperatures in a timely manner. This meant that warm meals were paid a medical service fee, which was unique in the world. The 1994 amendments included the abolition of the "Standard Meal System" and its replacement by the "Inpatient Meal Treatment System". In addition, the "additional fee for a special management meal" was renamed [the/an] "additional fee for special management", and the "additional fee for a patient cafeteria" and "additional fee for a select menu" were newly approved. The "additional fee for special management" is granted when a meal is provided to a patient under the guidance of a dietitian and meets certain conditions, such as being timely and at the right temperature. "Timely" means that dinner is served after 6:00 p.m., and "proper temperature" means that meals are served at a temperature maintained by means of heat-retention/cold-retention, heat-retention trays, heat-retention tableware, and the cafeteria.

The "additional charge for patient cafeteria" means that the hospital has a cafeteria and the floor space meets the condition of 0.5 square meters or more per hospital bed. The "additional charge for select menu" can be calculated when multiple menus that patients can choose from are provided for two main meals a day. Later, when timely provision of appropriate temperature and select menus became common, these add-ons were abolished, and the financial resources were used for clinical nutrition management in wards.

In 1994, the Asian Conference of Nutrition was held in Kuala Lumpur of in Malaysia and a symposium there focused on the computerization of nutrition services. I introduced examples of the computerization of nutrition services based on our own experience, and presented the principle of "software in front and hardware in back". In other words, interpersonal service for people was humanized, while the behind-the-scenes work was thoroughly streamlined. Therefore, dietitians/nutrition professionals should not be afraid of computers, but we should actively use them to improve the quality of our own work. After the presentation, the audience applauded loudly and asked many questions, and I remember being so moved that I got goose bumps.

At that time, the idea of creating an "Asian Congress of Dietetics: ACD" was floated by the gathered speakers. After a series of meetings, in 1994, First Asian Federation of Dietetic Associations; AFDA was established to develop research in practical nutrition in Asia (Photo 6.2).

The challenge of making safe, tasty, and hospital meals took several years of tireless work, and was unimaginably difficult. At the time, it was introduced as "The Challenge of Hospital Meals" on NHK news, and a video was made. It became a model for hospital food that patients can eat easily and comfortably. We were able to

ORGANIZING COMMITTEE
THE FIRST ASIAN CONFERENCE ON DIETETICS
JAKARTA, OCTOBER 2-5, 1994

We therefore resolve to form
The Asian Federation of Dietetic Associations

Signed in Jakarta, Indonesia
October 5, 1994

1. Hongkong Nutrition Association

Warren T.K. Lee

2. Indonesia Nutrition Association

Benny A. Kodyat

3. Japan Dietetic Association

Toji Nakamura

4. Korean Dietetic Association

Suh, Eun Kyung

5. Malaysian Dietitian Association

Fatimah Arshad

6. Nutritionist-Dietitians Association
of The Phillipines

L.N. Paniasigui

7. Singapore Nutrition and Dietetics
Association

Lynn Alexander

8. Taipei Dietitians Associations

Chwang, Leh-Chii

9. Thailand Dietitians Society

Rujira Sammasud

Photo 6.2   Declaration of The First Asian Federation of Dietetic Associations; AFDA at the first Asia Congress of Dietetics in Jakarta in 1994

do this because of the understanding and cooperation of Dr. Yukiko Kawashima et al. dietitians in the hospital and MEFOS Co., Ltd a food services.

6.5 Future Hospital Meals and Clinical Nutrition Management

The primary characteristic of medical care in Japan is that it is operated on the basis of the "National Health Insurance System" that began in 1961 (Showa 36). This system allows all Japanese citizens to receive medical treatment on an equal basis, and is financed by insurance premiums, taxes, and co-payments. Conventionally, medical service fees are paid by insurance and taxation by subtracting the patient's co-payment from the amount of the added fee, which is determined by the public price (one point = 10 yen) for each medical treatment, such as the fee for initial examination, examination, medicine, and guidance. This method is called the "piece-rate payment system" and has the advantage that medical institutions can determine the content of medical treatment without considering the cost. However, this method has also had the problem that it is difficult to control the increase in medical costs because hospitals earn more money if they provide high-cost medical care.

Therefore, a "Comprehensive Flat-Rate Payment System" is currently being implemented in Japan. The flat-rate payment system is a system in which the government sets an average price in advance for classified diagnostic groups, and medical institutions receive a fixed amount regardless of the cost of the procedure. This method makes medical care more cost-conscious and helps to control the total cost of medical care, because if medical care is more expensive than necessary, a deficit will result.

6.5.1 Controlling Medical Costs Through Clinical Nutrition Management

Hospital malnutrition is caused by loss of appetite, taste, digestion, absorption of nutrients, and metabolic changes. Malnutrition decreases the QOL of patients, reduces the therapeutic effect of drugs and surgery, increases the length of hospital stay, and ultimately increases the cost of medical care. At present, the importance of clinical nutrition management using hospital meals, enteral nutrition, intravenous nutrition, etc., has come to be emphasized because it is now understood that improving the nutritional status of injured and sick patients can increase the effectiveness of treatment and reduce medical costs. Clinical nutrition management has been evaluated as a method to bring about cheaper medical care, especially when the

"Fixed Payment System" is fully introduced, because the more medical institutions control medical costs, the more income the hospital will have.

In recent years, as patients with chronic diseases have become older, new challenges have arisen in dietary therapy.

First, the elderly have multiple diseases and complex disorders of various tissues and organs. Conventional dietary therapy for specific metabolic disorders is less effective. In other words, because there are multiple chronic diseases that cannot be cured completely, and because each condition is related to the others and exacerbation progresses, it is necessary to determine the priority of treatment and the appropriate amount of energy and nutrients based on a comprehensive understanding of the state of the whole body.

Second, diseases of undernutrition such as emaciation, anemia, sarcopenia, hypoalbuminemia, osteoporosis, and bone fractures are emerging among the elderly with diabetes and kidney disease. Even in the absence of disease, many elderly patients become frail. As they age, their appetite and taste buds decrease, their ability to chew and swallow decreases, and their ability to synthesize and break down nutrients decreases, meaning that they need more time to recover from disease. Diet therapy increases the risk of malnutrition if implemented for a long period of time because it forces people to eat an unbalanced diet compared to the diet of a healthy person.

6.5.2   Hospital Meals and the Way Out

Thus, as diet therapy becomes more diverse and individualized, the operation of hospital meal services is currently in a critical situation. The reasons for this are: ① A shortage of human resources to take charge of meal service operations, ② Meal service management is becoming difficult due to soaring labor costs, meal service material costs, and consumption taxes. ③ Hospital management itself is deteriorating, and ④ The number of people eating meals is decreasing due to a decrease in the number of hospital stays and the promotion of home medical care. Some meal service companies have withdrawn from the hospital meal business and will not accept orders from hospitals.

In the meantime, the following measures should be considered: securing of labor force including foreigners, improvement of the environment for food preparation workers including wages, and rationalization and simplification of operations. As a medium- to long-term plan, the following can be considered.

1. **Thorough rationalization of meal service operations**

We aim to streamline operations by constructing a food service system that is as rational as possible and by utilizing IT, AI, and robots for administrative and cooking tasks. In recent years, advances in cooking technology have led to research on the rationalization of food service operations, including the introduction of various processed foods and central kitchen systems. In this case, an important issue is how to incorporate the individuality of each patient's characteristics into the system.

2. Hospital meals and Clinical Nutrition Management

In promoting the rationalization of meal service operations, one thing that must not be forgotten is the response to the clinical diversity and individuality of the target population, which is the primary important characteristic of hospital meals. Specifically, it is a question of the linkage between clinical nutrition management through inpatient nutrition management and NST, which medical institutions have been actively working on in recent years, and hospital meals, which have been operated as a group meal facility. I believe that the key to both of these is the collaboration of dietitians with ward operations.

It is important for dietitians to understand and assess the nutritional status of the patients in the wards, to make a nutritional management plan, to monitor the nutritional status of the patients, and to communicate the information about the patients' preferences and eating conditions to the office and kitchen in a rational, prompt, and reliable manner so that this can be reflected in the meals. If a patient is left malnourished, recovery from surgery will be poor, the effect of medication will be poor, and nosocomial infections will increase due to lowered immunity, which will reduce the economic efficiency of medical care and increase the risk of medical safety management. Since nutritional management is already included in the basic hospitalization fee and the fee for nutritional guidance is double the previous amount, the assignment of dietitians to wards is now possible from the viewpoint of personnel costs.

Meanwhile, as co-payments for meals increase, patients are complaining more and more about the food and nutritional management. Some hospitals are beginning to receive complaints that patients have starved to death not because of illness but because of poor nutritional management. If dietitians were to spend time with patients, listen to their complaints about food, and deal with these issues, the problems could be solved and services could be improved. In other words, if the rationalization of food service is not based on the premise that a dietitian will be stationed in the ward, it will simply mean cutting corners on food service, and if forced, will cause new problems in safety management.

6.6   Characteristics of Diet Therapy by Disease

The following is a list of typical diseases requiring nutrition and diet therapy and their characteristics.

1. Diabetes

Diabetes is a disease that causes hyperglycemia due to insufficient or decreased action of insulin secreted by the pancreas, and over a long period of time leads to complications such as arteriosclerosis, renal disorders, retinopathy, and neurological symptoms due to abnormal metabolism of carbohydrates and lipids. The basis of dietary therapy is to normalize various metabolisms, mainly carbohydrate metabolism, as much as possible and to prevent complications; therefore, a low energy and low carbohydrate diet is implemented. Since an abnormal increase in postprandial blood glucose increases the risk of cardiovascular mortality, the glycemic index (GI) is used, and the increase in GI can be controlled by consuming soluble dietary fiber, fat, protein, vinegar, and milk and dairy products.

2. Dyslipidemia

Dyslipidemia is a disease in which cholesterol and triacylglycerol in the blood are in an abnormal state, triggering atherosclerosis. Diet therapy involves restriction of energy intake and carbohydrates as well as adjustment of lipid content. When obesity is a complication, improvement of obesity is the first priority. If total serum cholesterol and LDL cholesterol are high, saturated fatty acids should be restricted and polyunsaturated fatty acids should be ingested. However, excessive intake of polyunsaturated fatty acids also lowers HDL cholesterol and makes it susceptible to oxidative reactions, so when HDL cholesterol is low, oil containing a large amount of oleic acid, a monounsaturated fatty acid, should be used. A negative correlation has been observed between the intake of n-3 polyunsaturated fatty acids, which are abundant in fish and shellfish, and mortality from coronary events and myocardial infarction. These oils and fats have been shown to lower triacylglycerol, lower blood pressure, inhibit platelet aggregation, and improve endothelial function.

3. Hyperuricemia, Gout

Hyperuricemia is a disease in which the uric acid concentration in the blood is abnormally high, and gout is a condition in which uric acid crystallizes into urate, which accumulates in the joints and causes acute arthritis. The basic diet therapy controls the production of uric acid; overeating, obesity, high-purine and high-protein diets, and heavy alcohol consumption should be avoided. In order to increase the excretion of uric acid, adequate hydration should be provided. Patients should refrain from eating foods with extremely high purine content.

4. Hypertension

For patients with hypertension complicated by obesity, weight loss should be given priority, and a thoroughly low-sodium diet should be adopted. To reduce salt, use patients should substitute citrus fruits and spices, and use low-sodium foods such as low-sodium soy sauce and miso to make food more palatable. In addition, seafood, soybean products, milk and dairy products, vegetables, fruits, and seaweed should be actively consumed, and protein, dietary fiber, calcium, and calcium should be actively taken.

5. Chronic kidney disease (CKD)

Chronic kidney disease (CKD) is characterized by findings indicative of renal disease, such as positive urine protein, or a decline in renal function that lasts for more than 3 months. Diet therapy depends on the stage of CKD and is based on

Table 6.1 Significance of diet therapy (oral nutrition)

1	It is the most physiologically natural nourishment method and does not require special equipment
2	The quantity and variety of the food supplied are plentiful and less restrictive
3	The appetite and taste buds are satisfied, and mental satisfaction from the feeling of fullness is easily obtained
4	The endocrine and nervous systems are easily modulated by diet
5	The passage of food through the oral cavity is the immediate initiator of digestion, absorption, and metabolism
6	Unknown nutrients and active ingredients contained in food can be consumed

protein and salt regulation with an adequate energy supply. If obesity is present at any stage, weight loss is used to prevent worsening of CKD, and if hypertension is present, salt should be limited to less than 6 g/day. The stage of renal function decline is diagnosed by the glomerular filtration rate (GFR), and protein intake should be limited according to the stage. In the case of hypercalcemia, intake of potassium should be limited. If protein is severely restricted, it will be difficult to plan menus and prepare food on a daily basis, so low-protein foods for the sick should be utilized and their effect on nutritional status should be considered.

6. Surgery

Surgery is a major invasion of the organism and increases nutritional requirements. In addition, it is necessary to pay attention to the change in nutritional status because the amount of intake decreases due to diseases and disorders. This is especially evident in surgery for digestive diseases. The nutritional status of the patient should be improved before surgery, and depending on the type of disease, appropriate diet therapy and nutritional supplementation are required during and after surgery. The physiological risk increases as the degree of feeding, enteral feeding, and intravenous feeding increases, and it is necessary to try to rely on the oral diet as much as possible (Table 6.1).

If oral intake is inadequate, enteral feeds are often added. The method of nutritional supplementation and the amount of nutrition to be administered are determined by monitoring the patient's body weight, intake, digestion and absorption, and whether or not there is an increase in the amount needed. In general, after surgery, in consideration of the burden on the digestive tract the patient should start with a fasting diet, followed by a liquid diet, a 30% porridge diet, a 50% porridge diet, a full porridge diet, and then a regular diet, and the schedule should be constantly monitored. In order to lighten the burden from any one meal, frequent meals with snacks in between should be considered.

Bibliography

Endo M (1994) Enteral nutrition. Medicina 31(6):1154–1158
Iwasa M (1995) Shohei Ogoshi: recent trends in enteral nutrition. Igaku no Ayumi 173(5):479–483

Kondo K (2009) In: Nakamura T (ed) Clinical nutrition II: diseases and nutrition, 2nd edn. Dai-ichi Shuppan

Nakamura T (1998) Necessary techniques and systems for nutritional management. Nutr Assess Treat 15:9–14

Nakamura T (ed) (2017) The 3rd edition of essential diet therapy. Ishiyaku Publishers, Inc

Nakamura T et al (2008) Academic study on the effectiveness of dietary therapy in obesity, diabetes, kidney disease, and hypertension, Ministry of Health, Labour and Welfare of Japan: "Administrative Research for the Creation of Standards for Ensuring the Safety of Health Food," Summary Report 2007, pp 41–88

Nakamura T, Komatsu T, Sugiyama M, Kawashima Y (eds) (2020) 3rd revised edition clinical nutrition. Nankodo

Suzuki H (2015) In: Nakamura T (ed) Revised clinical nutrition II. Kenpaku-sha

Suzuki H (2016) In: Nakamura T (ed) Revised clinical nutrition I. Kenpaku-sha

Chapter 7
Nutrition for a 100 Year Life

Abstract The elderly loses or decline in various physical abilities, suffer from multiple chronic diseases, and have a higher risk of dying. However, even under these circumstances, they can lead independent daily lives and live happily by utilizing their remaining physical and mental functions. Furthermore, even if a person becomes ill and has a disability, it is still possible to be a healthy elderly person. and extending healthy life expectancy means increasing the number of such elderly people.

Energy-restricted diets for adults were effective in preventing obesity and metabolic syndrome, but not for children or the elderly. Energy restriction decreases bone density and increases the risk of sarcopenia and frailty. It is necessary to change gears in the life stage.

Keywords Healthy life expectancy · Energy-restricted diets · Sarcopenia · Frailty · Weakness · Nutrition Care Stations

7.1 What Is the Extension of Healthy Life Expectancy?

Throughout history, people have wished for health and longevity. This desire is especially strong among the powerful and the rich, and there are many stories from various countries about the search for food and medicine that will would grant health. Immortality, health and longevity. However, unfortunately, no one has yet fulfilled this wish. As long as human beings are living beings, the decline of physiological functions and the end of life due to aging are inevitable realities that come to all of us. In a sense, the search for food and medicine that will give us immortality is a challenge to our fate as living beings, and there is no solution yet. On the other hand our health has improved compared to 100 years ago, and we are able to live an incredibly long life. In other words, immortality is a pipe dream, but there is a good chance that we can extend our healthy longevity, our so-called healthy life expectancy, by improving our health, preventing diseases and advancing medical care.

T. Nakamura, *Japan Nutrition*, https://doi.org/10.1007/978-981-16-6316-1_7

In 2016, Professors Lynda Gratton and Andrew Scott of the London Business School shocked the world with the publication of "LIFE SHIFT: 100-year life - living and working in an age of longevity - (Toyo Keizai Inc.)". In developed countries, one out of every two people born in 2007 will live to be 103 years old, or will have a "100 year life", and Gratton and Scott advocated life planning based on the premise of living for 100 years. The average in Japan is expected to continue to rise, with half of children born in 2007 expected to live to age 107, maintaining the world's longest life expectancy. In September 2017, the Japanese government launched the Council on the 100-Year Life Era, chaired by Prime Minister Abe, to discuss economic and social systems in a super-longevity society.

7.1.1 The Challenges of a 100 Year Life

Now, in this super-aged society, can we live a healthy and happy life? What should our pensions, health, medical care, welfare, nutrition and diet be like?

The WHO has released an innovative report on ageing societies, the World Report on Ageing and Health 2015. What is innovative about this report is that it does not convey a sense of doom and gloom about aging societies, as is commonly the case. The report states that "the elderly are not dependents", "the aging of society will lead to an increase in health care costs, but not as high as expected", "look to the future, not the past", "consider spending on the elderly as an investment, not a burdened" [not a burden] "the cost burden of medical care and nursing care for the elderly is overemphasized, and social contributions are underestimated", "investment should be made in policies that support the elderly as well as efforts to reduce costs".

7.1.2 New Perspectives on Health

It is true that the elderly lose or decline in various physical abilities, suffer from multiple chronic diseases, and have a higher risk of dying. However, even under these circumstances, they can lead independent daily lives and live happily if they prevent and treat diseases that are more likely to occur as they age and are able to utilize their remaining physical and mental functions. Furthermore, even if a person becomes ill and has a disability, it is still possible to be a healthy elderly person, and extending healthy life expectancy means increasing the number of such elderly people.

There is another important aspect to the issue. It is health in the context of "extending healthy life expectancy" for the elderly, which is based on a different view of health than that of "health promotion", which aims to prevent the onset of disease. In other words, just as no one would call a Paralympic athlete who has lost a limb unhealthy, it is advocated that even if a person has cancer, diabetes, heart disease, kidney disease, etc., he or she should be an energetic elderly person who can do housework, hold a job, and engage in hobbies and volunteer activities. In short, we should aim for a society with a long life expectancy, where both the sick and the disabled can live independently and happily.

7.2 Physiological Changes and Nutrition in the Elderly

With aging, the functions of the whole body decline and the ability to adapt physically, mentally, and to the environment decreases. The changes vary from organ to organ, from person to person and from environment to environment. The physical changes include loss of height, atrophic changes in the intervertebral discs, flattening of the vertebrae, folding of the spine and lower limbs, weight loss, dryness of the skin, and loss of teeth. In terms of motor functions, movements become slow and unstable, and muscular strength and endurance decline. The decrease in muscle mass, which is a component of the body, and the decrease in water storage due to emaciation cause dehydration, and the decrease in bone mass makes the body prone to osteoporosis.

With aging, most organs atrophy, and physiological functions decline overall. However, differences occur among different organs. For example, cardiac output, which is a circulatory function, declines, while narrowing of the lumen of blood vessels and increase in resistance of peripheral blood vessels occur, as well as atrophy and loss of elasticity of the lungs. Among digestive functions, dryness of the oral cavity, decreased secretion of saliva, gastric juice/bile, and pancreatic juice, decreased masticatory function, decreased swallowing reflex, decreased contractility of peristalsis in the esophagus, and further decreased peristalsis in the intestines all occur, resulting in an overall decrease in the digestive and absorption functions.

In addition, decreases in the number of tongue papillae and unguent buds and in taste cell function cause a reduction in the sense of taste, and changes in taste are also observed due to changes in the sense of temperature and erosion pressure of the tongue and oral mucosa. In terms of hematopoietic functions, impairments in red blood cells, hematocrit level, hemoglobin level, serum iron, and iron-binding capacity occur.

As for mental functions, verbal ability, reasoning, and insight are relatively preserved, but nonverbal abilities such as intelligence efficiency, learning efficiency, memorization and recall decline.

7.2.1 Causes of Malnutrition in the Elderly

In the midst of these physiological changes, the elderly are more likely to suffer from malnutrition. The reason for this is that overall dietary intake declines in old age due to changes in taste, loss of teeth, and decreases in masticatory strength and saliva volume, in motor functions of chewing and swallowing, and in saliva due to increases in medication and in digestive enzyme activity (Table 7.1). This is also related to a change in the content of the diet from meat dishes to lighter dishes such as seafood and vegetables, and a decrease in the intake of fats and oils, meat, milk and dairy products, and eggs. This leads to a lack of energy intake, protein, fat, vitamins and minerals.

Reduced food intake tends to result in weight loss, especially lean body mass (LBM), as well as in loss of muscle strength and of intercellular water. As muscle strength and physical activity decline, a appetite weakens, and at the same time, muscle mass is lost, causing a lowering of basal metabolism, leading to a negative spiral in which the amount of energy consumed decreases. Ultimately, low nutrition results in lessened vitality, increased fatigue, and a lower quality of life (QOL). Such undernutrition in the elderly increases the risk of developing deficiency diseases such as emaciation, energy and protein deficiency, sarcopenia, iron deficiency anemia, and osteoporosis due to calcium deficiency.

In addition, even if such nutritional deficiencies do not develop, the deficient state will lead to the development of various complaints.

7.3   The Risk of Eating in Moderation for Elderly People

Many Japanese people believe that a moderate eating diet is effective for good health and longevity. Is this really true?

In nutritional terms, a moderate eating diet implies an "energy-restricted meal". In 2009, the University of Wisconsin (UW) in the U.S. announced the longevity effects of an energy-restricted diet on rhesus macaques, after raising them for many years (Fig. 7.1). Both males and females lived longer when maintained on an energy-restricted diet. This result led many researchers to believe that a moderate eating diet is a healthy diet for longevity. However, in 2012, the National Institute on Aging (NIA) in the U.S. reported that an energy-restricted diet was not effective in reducing mortality and extending life span in rhesus monkeys in an experiment similar to that conducted at UW. Why did they get different results from nearly identical experiments? The two groups argued constantly for 5 years. But, deciding not to continue the controversy any longer, the two groups came to the same table, examined the causes in detail, and published a jointly-authored report in 2017.

Table 7.1 Factors causing decrease in food intake

1	Aging	Loss of appetite, loss of sense of smell and taste
2	Diseases	Mastication and swallowing disorders, digestive disorders, inflammation and cancer, metabolic disorders, side effects of drugs
3	Psychiatric and psychological	Cognitive dysfunction, depression, fear of aspiration and choking
4	Society	Living alone, lack of care, loneliness, poverty
5	Others	Incompatible eating patterns, excessive reaction to obesity and lifestyle-related diseases, faulty nutrition and diet knowledge

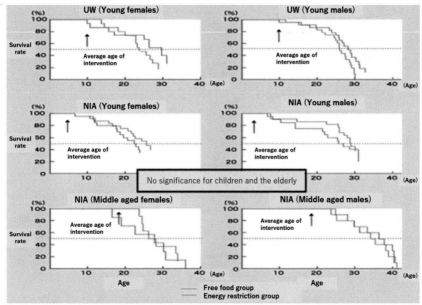

UV study: Survival rates were high for both males and females in young
　　　　　age group 7-15 years (21-30 years in humans).
NIA study: Males have no difference at young ages 1-5 years (human ages
　　　　　3-15 years), females have lower survival rates.
　　　　　No difference between males and females at 16-23 years (48-69
　　　　　years in humans).

Fig. 7.1 Benefits of an energy-restricted diet

7.3.1 *Pitfalls of Energy Restriction Research*

The reason for the different results was simple. It was the difference in age at the start
of the energy-restricted diet intervention. That is, the UW study included monkeys
aged 7–15 years (21–45 in human years), while the NIA study included monkeys
aged 1–5 years (3–15) and 16–23 years (48–69), respectively. In other words,
energy-restricted diets for adults were effective in preventing obesity and metabolic
syndrome in middle-aged and older adults and were associated with longer life, but
not for children or the elderly. In addition, the monkeys that responded well to the
energy-restricted diet showed a decrease in bone density, indicating that they had an
increased risk of osteoporosis despite their longevity. In the end, a moderate eating
diet in the elderly does not lead to longevity, and energy restriction decreases bone
density and increases the risk of requiring nursing care (Fig.7.2).

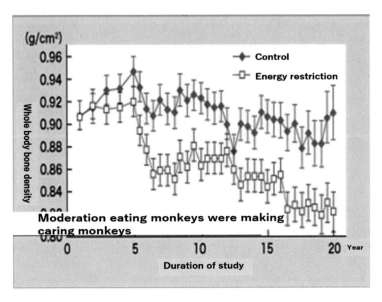

Fig. 7.2 Loss of bone density due to long-term energy restriction. (UW Study (Age (Dordor) 34:1133–1143, 2012))

7.4   Sarcopenia and Frailty

Sarcopenia and frailty are attracting attention as conditions that can lead to the need for nursing care and reduce life expectancy Sarcopenia refers to the decline in skeletal muscle mass due to aging, with a secondary decline in muscle strength and aerobic capacity. Loss of muscle mass is an essential factor, and sarcopenia is diagnosed if either muscle strength or physical ability deteriorates. On the other hand, "frailty," which is translated into Japanese as "weakness", not only refers to a decline in muscle mass and function, but also includes mental frailty, which leads to cognitive loss and depression, and social frailty, in turn leading to withdrawal and decreased communication with others. In other words, while sarcopenia is a disease in which the main symptom is a decrease in muscle mass in all age groups, frailty is a syndrome in which the functions of the entire body decline with aging.

7.4.1   Causes of Sarcopenia and Frailty

A common factor that triggers this condition is undernutrition, which is a lack of energy and protein. Undernutrition leads to an increase in fatigue, as well as decreases in vitality, walking speed due to muscle weakness, and activity, which increase the risk of both sarcopenia and frailty, and increase the risk of nursing care.

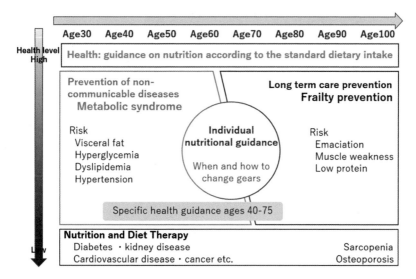

Fig. 7.3 Shifting from metabolic syndrome prevention to frailty prevention

In recent years, frailty has received particular attention as we face an aging society. A study has been published that observed men in Scandinavia for a long period of time from the 1970s to 2007. When long-term changes in BMI and the onset of frailty were examined in four groups, there was no relationship between the normal weight-unchanged group, the consistently overweight group, and the weight gain group, but there was a significant increase in the onset of frailty in the weight loss group.

What is important in preventing frailty is prevention of emaciation and muscle loss due to lack of energy and protein, which ultimately means eating well. On the other hand, in order to prevent lifestyle-related diseases, which are non-communicable chronic diseases, it is important for middle-aged and older people to take active measures against metabolic syndrome and control their weight by eating an adequate amount of food. In other words, in order to prevent age-related frailty, it is necessary to eat well and not lose weight, and when one reaches a certain age, it is necessary to change gears (Fig.7.3).

7.4.2   How to Become an Independent Elderly Person

Akihiko Kitamura et al. of the Tokyo Metropolitan Institute of Health and Longevity Sciences observed the incidence of loss of independence, need for long-term care, and death from frailty and metabolic syndrome categories as outcomes in subjects aged 65 years or older who received medical examinations (Table 7.2). The results showed that the more advanced the frailty, the higher the incidence of any of the outcomes. The incidence of loss of independence over a 7-year period was about

Table 7.2 Incidence of self-loss, long-term care needs and death by frailty category (7 years)

		Frailty category		
		Non-frailty	Pre-frailty	Frailty
Male	\<Average age\>	\<69.5\>	\<71.1\>	\<74.8\>
	Loss of self	22.8	42.9(1.9)	110.4 (4.9)
	Long-term care required (Including support required)	10.7	24.4(2.3)	77.3(7.2)
	Long-term care required (level 2 or more)	5.0	11.1(2.2)	42.8(8.6)
	All deaths	29.5	53.6(1.8)	124.7 (6.1)
	Deaths from cardiovascular disease	2.9	9.3(3.2)	38.4 (13.3)
Female	\<Average age\>	\<68.7\>	\<70.9\>	\<75.7\>
	Loss of self	13.6	32.9(2.4)	90.8(6.7)
	Long-term care required (including support required)	11.9	26.7(2.3)	77.4(6.5)
	Long-term care required (2 or more)	5.9	8.8(1.5)	32.0(5.4)
	All deaths	5.3	20.9(0.3)	58.1 (11.0)
	Deaths from cardiovascular disease	1.8	6.2(3.5)	20.3 (11.3)

twice as high in the pre-frailty group and about five times as high in the frail group as in the non-frailty group for men, and about 2.5 times as high in the pre-frailty group and about 6.5 times as high in the frail group for women. The age-adjusted hazard ratios for the outcomes were higher in the pre-frailty group than in the post-frailty group, and the risk ratios were three to four times higher in both groups, indicating that it is more effective to detect and address frailty early in old age.

On the other hand, when we look at the metabolic syndrome category, the incidence of outcomes in the metabolic syndrome group and the non-metabolic syndrome group was not related to any of the items, and it was understood that metabolic syndrome measures had no effect. In other words, at the age of 65 years or older, those with a BMI of 25 or higher should continue to lose weight by eating a moderate diet, but the aging of patients with chronic diseases will require new measures.

For all outcomes, the frailty group was significantly higher than the "non- frailty group". Loss of self was five times higher. Loss of self is the first Certification of Needed Long-Term Care or death before Certification.

New problems are beginning to arise for elderly people who already have chronic diseases. For example, elderly diabetic patients are more likely than non-diabetic patients to suffer from sarcopenia, cognitive decline, ADL decline, falls, bone fractures, and other geriatric syndromes and frailty. A hemoglobin A1c of 8.0% or higher is associated with various complications of diabetes, while a hemoglobin A1c of less than 7.0% is associated with bone fractures, falls, and frailty. Add different Countries are considering setting higher hemoglobin A1c targets in elderly diabetics than in the general diabetic population.

7.4.3   Response to the Individualization of Nutrition and Diet Therapy

Against this background, in September 2019, The Japan Diabetes Society published its Diabetes Care Guidelines 2019. While the method of calculating total daily energy intake is body weight (kg) × energy coefficient (Kcal/kg), as in the past, in the Guidelines body weight was changed from standard weight to target weight, and the energy coefficient was made flexible, taking into account not only the physical activity level but also the pathological level. In other words, the entire diet therapy should be individualized, and at the same time, it should be tailored to the individual pathological and nutritional conditions of elderly patients, especially those at high risk of obesity, underweight, malnutrition and frailty. Specifically, the guideline for the target weight (kg) should be set individually within the range of height $(m)^2 \times 22$ for those under 65 years old and height $(m)^2 \times 22$–25 for those 65 years old and over; especially for late-stage elderly patients aged 75 years old and over, it should be determined based on the current weight and assessment of frailty, ADL decline, concomitant diseases, body composition, shortening of height, eating status and metabolic status. The decision should be made as appropriate based on the preceding factors. The amount of physical activity is also expressed as a coefficient and is chosen from the following three levels (Table 7.3).

With the ageing of the population, diet therapy has become more individualized and methods of nutritional management more complex. Nutritional status, therapies, medications, genetic make-up, nutritional support methods, as well as community and family environment, relationships, learning ability and financial situation, must all be considered to determine the most appropriate nutritional and dietary treatment for the individual. To achieve this, it is necessary to create a place where people can get advice from a dietitian in close proximity where they live, and The Japan Dietetic

Table 7.3 Energy coefficients (Kcal/kg)*[1] by physical activity level and disease state

1. Light exertion (mostly static activity in a seated position): 25–30
2. Normal exertion (mostly sedentary, but includes commuting, housework, and light exercise): 30–35
3. Heavy exertion (heavy work, active exercise habit): 35~
A coefficient larger than the physical activity level can be set for frailty prevention in the elderly, and a coefficient smaller than the physical activity level can be set for weight loss in the obese. In either case, if there is a discrepancy between the target weight and the current weight, the coefficient should be set flexibly with reference to the above

Association launched the "Nutrition Care Stations Certification System" in 2018 (Heisei 30). The Nutrition Care Station offers a wide range of services, including nutritional dietary counseling, specific health guidance, and seminars and workshops for local residents as well as municipalities, health insurance associations, private companies, and medical institutions.

Also, in hospitals, registered dietitians have been becoming to taken a role in promoting nutritional management in intensive care units for early recovery of patients.

7.5 Nutritional Dietary Care for the Elderly at Home

In the elderly, diseases become more complex, and at the same time, the decline in physical and mental functions affects their lives. Therefore, it is necessary both to improve nutrition to enhance the intrinsic functional capacity of the individual and to improve their environment to enable an independent diet. Low nutritional status causes weight loss as well as physiological changes such as decreased body temperature and pulse rate, decreased physical strength, edema, and decreased vision and hearing. Furthermore, it is noteworthy that mental changes in the elderly, such as lower concentration and attention, depression, irritability, lethargy, and hysteria, occur. In other words, low nutrition causes a decline in mental and physical functions, regardless of the disease. Nutritional management is necessary to maintain and improve the QOL of the elderly as well as to treat diseases. In order to do this, a nutritional assessment must be conducted to evaluate and determine the nutritional status of the subject, a nutritional care plan must be developed based on this assessment, and dietary therapy and nutritional supplementation must be monitored and reevaluated. This will help to maintain and improve the physical and mental functions of the elderly and extend their healthy life expectancy.

Human beings always wish to be healthy and live a normal life. However, in order to maintain a healthy life expectancy while feeling a sense of well-being, people often wish to live in their own homes as long as possible and wish to have nutritional management at home.

7.5.1 Key Points on Improving Nutrition for the Elderly at Home

The following are the key points to improve nutrition for the elderly at home:

① Hyponutrition in the elderly is seen as a generalized malnutrition, so certain nutrients and foods are not the solution; we need to make sure that the elderly are not deficient in all nutrients. Energy and protein, of course, but also vitamins and minerals lacking. In the case of a decrease in food intake, it is necessary to know why the person is unable to eat, and to improve these factors to the extent possible. Specifically, it is important to use vinegar, spices, and highly palatable foods and preparations that increase appetite in moderation, as well as to control the temperature of food, the atmosphere of the table and room, and smells. Furthermore, it is important to eat together as much as possible and not to talk about unpleasant topics during meals. In addition, if chewing and swallowing are difficult, it is also a good idea to use special-purpose foods like thick liquid foods and foods that are easy to chew and swallow.

② Human relations with the subject, family members, housemates, and people in the community surrounding the home are important; it is also necessary to understand the relationship and characteristics of the community, as well as the environmental conditions that support meals at home, such as convenience stores, supermarkets, food stores, and even restaurants and cafeterias.

③ The capacity and support of the target population themselves, their families and communities for knowledge of and skills in nutrition management also need to be investigated as part of the support system. Once these are ascertained, a plan of care is constructed based on the results of the assessment of the subject as to how the actual diet and enteral/intravenous nutrition will be implemented.

In the elderly, even gender, age, and health and disease status are the same, we should realize in mind that there is no standardized diet or diet therapy. It is necessary for physicians, registered dietitians, nurses, pharmacists, public health nurses, physiotherapists, occupational therapists, and others to work together to create and implement individualized and comprehensive nutritional management methods that take into account the degree of illness, diversity, and risk of health problems, as well as lifestyle and medications, of individual subjects.

Bibliography

Akihiko K, Shoji S, Masaru T et al (2017) Medium- and long-term effects of frailty and metabolic syndrome in old age on loss of independence defined using information on nursing care needs: the Kusatsu-cho study. Jpn J Public Health 64(10):593–606

Beard JR, Officer A, Cassels A (2015) World report on ageing and health. World Health Organization, Geneva

Colman RJ, Anderson RM, Johnson SC et al (2009) Caloric restriction delays disease onset and mortality in rhesus monkeys. Science 325:201–204

Keys A, Brožek J, Henschel A, Mickelsen O, Taylor HL (1950) The biology of human starvation (2 volumes). University of Minnesota Press

Mattison JA, Roth GS, Beasley TM, Tilmont EM et al (2009) Impact of caloric restriction on health and survival in rhesus monkeys from the NIA study. Nature 489:318–321

Mattison JA, Colman RJ, Beasley TM et al (2017) Caloric restriction improves health and survival of rhesus monkeys. Nat Commun 8:14063

Strandberg TE, Stenholm S, Strandberg AY et al (2013) The "obesity paradox," frailty, disability, and mortality in older men: a prospective, longitudinal cohort study. Am J Epidemiol 178 (9):1452–1460

Zaslavsky O, Walker RL, Crane PK et al (2016) Glucose levels and risk of frailty. J Gerontol A Biol Sci Med Sci 71:1223–1229

Chapter 8
Nutrition in Japan from an International Perspective

Abstract International Congress of Dietetics: ICD was born in 1950 in Amsterdam, and was held every four years during the Olympic year. International Congress of Nutrition (ICN) which is a congress on basic research in nutrition, and the ICD which is a congress on practical research in nutrition and dietetics. We holded the 15th ICD in Yokohama (photo). What I wanted to convey to the world most at this congress was Japan nutrition and contribution to the world. The traditional Japanese diet was not healthy diets but low in energy, protein, fat, vitamins and minerals, many Japanese people suffered from nutritional deficiencies. Japan succeeded in improving its nutrition and became a country of longevity because the government drew up an excellent nutrition policy, dietitians and nutritionists worked hard to educate it, and the people believed in it and put it into practice. Intergovernmental Tourism Organization (JETRO) presented us with the "International Congress Attraction and Hosting Contribution Award" for the best management of an international congress held in 2008.

By hosting the ICD in 2008, we learned that we should teach people in other countries how to solve the malnutrition and to have a health and long life. Particularly in developing countries in Asia and Africa, where hunger and stunted growth of children due to low nutrition are still unresolved, some wealthy people are suffering from obesity and diabetes. We have contributed to the education and training of registered dietitian Hanoi Medical University in Vietnam.

Keywords International Congress of Dietetics: ICD · International Confederation of Dietetic Association: ICDA · International Congress of Nutrition: ICN · International Congress Attraction and Hosting Contribution Award · Hanoi Medical University · Japan International Cooperation Agency: JICA

T. Nakamura, *Japan Nutrition*, https://doi.org/10.1007/978-981-16-6316-1_8

8.1 "We Could Really Have an International Conference in Japan!"

The 15th International Congress of Dietetics (ICD 2008) was held at Pacifico Yokohama on September 8, 2008 (Heisei 20). The main theme of the Congress was "Collaboration and Cooperation among Dietitians around the World for the Health of Humankind". 8028 people from 57 countries participated in the Congress, and the main hall of the National Convention Hall, the largest in Japan, was filled with participants. The flags of the participating countries were lined up on the stage, and the hall was filled with unusual excitement when solemn music and a video introducing the significance and history of the ICD were played. A welcome party was held in the Marine Lobby of the Main Hall from 18:00 on the previous day. The number of participants was much larger than expected and the venue was filled with a lively welcome party with Japanese drums, jazz music, and even dancing.

8.1.1 Opening of the ICD in Japan

At 8:30 on the day of the conference, the curtain finally opened on the ICD in Japan. The words "Opening" were projected on a giant screen, and the stage was lined with ICDA (International Confederation of Dietetic Association) directors and officials from the host country. Behind them were the flags of the participating countries, in front of which the 57 national represen[t]ive were seated. Following the opening remarks by Motoko Sakamoto, ICDA Japan Representative Director, and Sandra Capra, ICDA President, I, as the Chair of the Organizing Committee of the host country, made a speech. In my address, I pointed out that there are problems of under-nutrition and over-nutrition in the world, and that in developing countries, there are problems of under-nutrition caused by poverty and over-nutrition among the wealthy due to rapid economic growth, and that in developed countries, although there is an abundance of food, a new type of under-nutrition is becoming a serious problem among the injured and sick, the elderly, and young women. I mentioned that face-to-face international exchanges among nutritionists and dietitians from different countries are necessary to solve these problems. In fact, the emblem for the conference was designed by Makoto Wada, a world-renowned illustrator. The message was that although we live in an age of global advances in communication technology, what is important is for human beings to communicate with each other face to face (Photo 8.1).

The opening ceremony address took about six months to write and was ready a month before the Congress began.

Photo 8.1 Poster of the 15th ICD (ICD 2008)

8.1.2 Opening Declaration with Ad-Libs

However, as the event approached, I began to feel inadequate. Not even a week before, not even the day before, could I find the words to express my strong feelings about the stage I had been dreaming of for about 30 years. I wondered if this was the kind of situation where one's thoughts and feelings exceed words.

On the morning of the opening ceremony, I made up my mind. I'll just go up to the podium and say what comes naturally to me at that moment. I decided that the words could be in Japanese or English.

One by one, I ascended the steps to the platform of the grand stage, and suddenly I saw the gazes of the participants filling the convention hall, all focused on me. "What a beautiful view!" I thought. My immediate words were, "Everyone. Please take a

Photo 8.2 The opening ceremony

look. We really had an international conference in Japan." The audience erupted in applause, and the sound gradually spread until it became a din sound that filled the hall.

When I finished, many people said to me, "With that one word, I understood all your feelings (Photo 8.2)."

The Minister of Health, Labor and Welfare, the Governor of Kanagawa Prefecture, and the Mayor of Yokohama gave congratulatory speeches.

The Full Range of ICD 2008 In the exhibition hall, there were 52 booths of Japanese companies and organizations, 5 booths of overseas companies, the Nutrition Care Station of the Japan Dietetic Association, and booths of 47 prefectures. In one corner of the exhibition hall, a kitchen stage was set up where 6 companies and organizations from Japan and overseas performed 9 demonstrations for 2 h each. On the 7th, a "Nutrition Walk 2008" was held and a thousand people walked the streets of Yokohama wearing badges with the logo "ICD 2008 Yokohama". An evening concert by the "Japan Philharmonic Orchestra" (at Yokohama Minato Mirai Hall) was held on the 8th, and a banquet was held (at the Pan Pacific Yokohama Bay Hotel Tokyu) on the 9th, and there were also fun events such as a tour of downtown Edo and Kamakura.

At the closing ceremony, Dr. Shigeru Yamamoto, Chair of the Executive Committee, expressed his gratitude to the participants, and Dr. Marsha Sharp, ICDA Canada Representative Director, explained the history of ICDA's development, organization, mission, and future activities.

8.2 Determined to Attract International Conferences

The "ICD 2008 Yokohama" was a great success. However, the road to it had not been an easy one. This congress was born in 1950 in Amsterdam, the Netherlands, and is held every four years during the Olympic year, mainly for nutritionists and nutrition scientists who are engaged in practical activities in nutrition. International congresses on nutrition include the International Congress of Nutrition (ICN), which is a conference on basic research in nutrition, and the ICD, which is a conference on practical research in nutrition and on the state of the profession. The 5th conference in Washington was attended by John F. Kennedy, the 10th in Paris by the then Mayor of Paris, Jacques Chirac, and the 13th in Edinburgh received a message from Queen Elizabeth II. It was a great honor to be able to hold such a prestigious international conference on nutrition in our country, and there was a long and arduous road to attracting such a conference.

8.2.1 The Hard Road to Attraction

It was at the time of the ICD in Sao Paulo in 2004 that I made up my mind to invite this international conference to Japan and began to dream of its realization. After the ICD board meeting, I went to the hotel with the president of the Japan Dietetic Association, Mr. Morikawa.

He said "Mr. Nakamura, the Board of Directors recommends that the ICD be held in Japan in eight years." There are many Japanese immigrants in Brazil and the overall atmosphere of the conference encouraged the decision to hold the conference in Japan, which was recommended by the then President of the Brazilian Dietetic Association.

"President Morikawa, if anyone should be the first in Asia to host the conference, it should be Japan. Let's do it!" It was a good idea, so I recommended it.

However, former president Morikawa's words were surprising.

"Unfortunately, the Japan Dietetic Association does not have the financial resources or language skills to hold an international conference."

I never forgot the sad face of former president Morikawa after that. At this time I began to think about inviting this international conference to Japan.

After the conference, I went to Machu Picchu, a World Heritage Site. I went down the mountain and headed for the station, but the train was stopped by a squall of heavy rain. When we were waiting in the train, we looked outside and saw a girl standing on the gravel on the track, soaking wet, and trying to offer us flowers. The rain was so heavy that we could not open the window. She held the flowers above her head, barefoot, and refused to leave the spot. I indicated with my hand that I could not buy them and that she should go now, but the girl would not move.

8.2.2   The Decision to Attract

Former president Morikawa and his wife, who were sitting in front of me, said, "We, too, were poor after the war and struggled in the same way. If you work hard, things will get better someday," he said over and over again. I just listened in silence. We were born Japanese and were able to escape from poverty due to economic development, but she was born in this poor village by not her own choice, and I felt that she was not responsible for this. This experience later became the starting point of my own passionate commitment to making an international contribution and the driving force behind my efforts to attract this international conference.

8.3   Red Roses on the Congress Bag

8.3.1   Defeated by the Philippines

Attracting ICD was not an easy task. The first official announcement of an invitation to Japan was made at the ICD General Assembly held in Paris in 1983. At that time, a rival unexpectedly appeared on the scene. It was Manila in the Philippines. At that, the Marcos regime had collapsed in the Philippines and the Aquino presidency was underway. She was the first female president, with outstanding international recognition and tenure, and was also interested in nutrition issues. The American Dietetic Association, which was the largest sponsor of the International Dietetic Association, also supported the Aquino administration. On the other hand, at that time, Japan was in the midst of high economic growth, "Japan is No. 1: Lessons for America" became a best seller, the Japanese economy was dominating the world, and Japanese people were walking around the world. We, too, believed that there was no way Japan could lose to the Philippines.

The result was that we lost by one vote.

I was not in charge of international affairs at the Japan Dietetic Association at the time, so I did not attend the delegate meeting where the election took place. According to what I heard from those who attended the meeting, our country brought many gifts and made a speech to introduce the development after the war, but the Philippine delegate said, "We have no money and cannot give you a lavish gift, but we sincerely welcome you." After the conference, the executive committee of the ICD at the time advised me that if Japan really wanted to attract the conference, it should make more international contributions, and I became the first Japanese person to be elected as a director of the International Confederation of Dietetic Association.

After that, the Board meetings were held in many parts of the world, and I spent my life traveling abroad. The greatest thing about being on the International Board was that I made three friends who would go on to become key players in global nutrition. They were President Eileen Mackay from the UK, President Sandra Capra from Australia and President Marsha Sharp from Canada. Eileen went on to become

President of The European Federation of the Associations of Dietitians, and Sandra and Marsha each became President of the ICDA. At that time, it is no exaggeration to say that the four of us, including myself, were deciding the direction of practical nutrition in the world, and the three of them called me "Teiji" affectionately.

8.3.2 Victory over Australia

In September 2000, the fateful moment arrived. The 13th ICD was held in Edinburgh, and at the delegates' meeting, it was officially announced that the ICD 2008 would be held in Japan. The rival was Sydney, Australia, where Sandra was the president. The Sydney Olympics were being held that year in 2000, and the whole world was focused on Sydney, putting Japan at a definite disadvantage. In our presentation, we talked about Japan's international contributions to date, the results of Japan's efforts to improve nutrition, and the fact that Yokohama is an international city and the site of the World Cup soccer finals, a popular event in Europe. The elections would be held in three days, and the two countries would be battling to attract more candidates.

Australia was a member of the British Commonwealth, and if these countries united, Japan would not have stood a chance. Australia's powerful president, Sandra, did a great job in gaining support; however, Japan was also keen to attract the attention of everyone attending the conference. We set up a booth in the style of a tea ceremony room, complete with red carpet, and handed out posters, pamphlets and other small items, as well as holding a green tea ceremony. Toward the end of the competition, Ms. Eileen, chairperson of the British, and Ms. Marsha, chairperson of the Canadian delegation, secretly whispered to me, "We will support you, Teiji" would be more idiomatic.

The next day, I attended the delegates' meeting with a nervous feeling. As I was about to enter the hall, I saw an unbelievable sight. The representatives of the countries related to the British Commonwealth were wearing red roses on their congress bags. It is said that the representatives of these countries got together and decided to support Japan and promised to wear a rose on their bags on the day of the election as a sign of their support.

In the hall, President Sandra was sitting right in front of me in the front row. The election result was an overwhelming victory for our country. Without looking back at me, she put her hand behind her back and said, "Congratulations". When I looked at her feet, I saw that she had champagne ready for herself. It seems that she had been confident. Confident that she would win and was planning to open the bottle on the spot to celebrate her victory. All the Japanese who attended the meeting in Edinburgh were delighted.

8.3.3 Busy Days of Preparation

There were eight years until Japan hosted the congress, but the difficulty of the preparations was beyond imagination. Each committee held many meetings over and over again. I was personally advised by a company that specializes in consulting for international conferences. It became clear to me that there was a lot of work to do in preparation. At the end of their presentation, I asked them, "What is the most important thing to make it successful?"

The answer was really quite simple.

"It's about money. How much money you can raise that will determine whether we succeed or not."

I, and the Japan Dietetic Association, had been hit with what we are most uncomfortable with. However, I understood that this was the most difficult thing to do, but that it was essential for success, and I was determined to become "money-hungry". At every turn, I bowed to the people concerned and asked them to help the company. I said to the members, "This international congress will not be a success unless the members work together as one. As a sign of this, each member is asked to support us with 10 yen per month" at the general meeting, and this was approved. We thought that if we did not show that we were willing to give money, others would not support us. As a result, the number of sponsors exceeded our expectation, and "Yakult Honsha", "Ajinomoto", "Otsuka Pharmaceutical" and "Kagome" became the gold sponsors at 10 million yen each, which gave us an impetus to raise further funds.

8.4 Hitting the Wall Many Times

During the meeting, there were many times when we could not come to an agreement, the program was not finalized, the speaker was not decided, and many other problems came up. Furthermore, we faced a big personal problem.

8.4.1 Overcoming Life Threats

In 2003 (Heisei 15), I moved from St. Marinna University Hospital to my current position at Kanagawa University of Human Services. This university was established in Yokosuka City as part of Kanagawa Prefecture's 21st Century Plan, with the aim of training human resources in healthcare and welfare to cope with an aging society. I participated in the preparations for the university two years before its opening date, and every day was like a storm, with abnormal days overlapping with the preparations for an international congress. In April 2003, the Kanagawa University of Human Services was opened with great success. However, after that, I fell ill and was hospitalized during the consecutive holidays in May. The initial diagnosis at the time of hospitalization was severe iron deficiency anemia, but without knowing the cause of the anemia, I was discharged from the hospital because my health had recovered for the time being. I repeatedly visited the outpatient clinic, but my condition did not improve, so I went to see a hematologist. On 12 March 2004, I was diagnosed with "malignant lymphoma". I was told that I had cancer. I immediately asked the doctor.

"Sir, please let me live until 2008."

There was no definite reply from the doctor.

I was then readmitted to the hospital and underwent radiation therapy. Since cancer informing patients of a cancer diagnosis was not yet common, I was told that I had an "intractable duodenal ulcer". When the former president of the Japan Dietetic Association, Pro. Fujisawa, and the former chairman of the board of directors, Mr. Hanamura, came to visit me, I decided to let them know the truth, so I turned up my T-shirt and showed them my stomach where the radiation would be applied, I remember the shocked looks on their faces. Fortunately, it seemed to be a type of radiation therapy that was effective, and I managed to make it to the international congress, and I am still alive today.

In 2004, the 14th ICD, co-hosted by the United States and Canada, was held in Chicago. As this was the last conference before the ICD was held in Yokohama, I started to prepare to invite people to participate in the "ICD Yokohama 2008" and to make a speech as the next host country. At that time, former Prime Minister Junichiro Koizumi was immensely popular not only in Japan but also internationally. Thinking that this was impossible, I asked the government if a welcome message from the Prime Minister could be played on video at the congress in Chicago. To my surprise, the government agreed.

8.4.2 The Then Prime Minister Koizumi's Quest for Perfection

On May 13, 2004, I went to the Prime Minister's residence with a film crew. In a prepared room, Prime Minister Koizumi entered with five guards and an expert who checked the English text. I thought it would only take one or two takes, maybe 20 min, but the prime minister made many, many revisions until he was satisfied. Each revision required time for adjustment, and each time I was able to talk with the prime minister alone. We chatted about our hometown of Yokosuka. We managed to finish the filming, but the next day we found ourselves in a serious situation. The Prime Minister's schedule after the filming had gone away, and it was reported newspapers "A Day in the Life of the Prime Minister" that some of his scheduled had been changed. I rushed to the Prime Minister's Office to apologize.

At the ICD General Assembly in Chicago, a video message from Prime Minister Koizumi was played in the hall immediately after my address concerning the next opening of the conference. The venue was filled with loud applause, and I received words of praise from the Directors of ICDA, who said, "Well done, Teiji."

Apart from such glamor on the surface, I also experienced a sad reality behind the scenes.

As the next host country of the conference, we were planning to have a lively display full of Japanese colors at the exhibition hall. Thinking that we would not be able to set it up it in time on the day of the conference, I went to the venue with the people in charge the day before, set up the display, and decorated it with excitement. The result was even better than we had expected, and everyone returned to the hotel satisfied.

8.4.3 Deep-Rooted Resentment

Early the next morning, I walked into the exhibition hall and was amazed.

The booth's display had been shattered. The five staff members in charge of the booth were at a loss for words and fought back tears of frustration as they worked to correct the damage, and managed to restore the appearance of the booth just before the doors opened. Of course, we didn't know who the culprit was, and we couldn't bring ourselves to complain about it to the organizers, so none of us ever spoke about it to the outside world after that. I remembered that someone once told me, "Be careful because some Americans still think that Japan is an enemy country." I then remembered an exhibit of posters that were put up during the Pacific War in a corner of the Smithsonian Institution in Washington. There was a poster that read "Never forget Pearl Harbor" in an uplifting message to crush the enemy, Japan. I learned once again that the scars of war remain not only in the time of those who were directly involved in the war, but also for generations to come.

War should never be fought, and people should get along well in the world.

8.5 Introduction of the History of Nutritional Improvement in Japan

In preparing for the ICD, there was something I wanted to share with the people of the world. It is the reason why the Japanese people have the longest life expectancy in the world. At the time, when I went abroad, there was a boom in Japanese food, and magazines and newspapers reported on it. Most of the articles were about the health benefits of traditional foods and dishes such as rice, sashimi, tempura, sushi, tofu, natto, and miso soup. I felt uncomfortable with such reports and thought that I had to let people in other countries know the real reason "why the Japanese people have the longest life expectancy in the world".

At this international congress, I organized a symposium entitled "Health and Nutrition Policy of Japan-Why do Japanese live long?"

8.5.1   From a Short-Life Country to a Long-Life Country

As one of the speakers, I told the following story.

From a nutritional point of view, the traditional Japanese diet was low in energy, protein, fat, vitamins and minerals. Before the Westernization of the diet, many Japanese people suffered from nutritional deficiencies. The neonatal mortality rate was high, children's growth and physique were poor, their resistance was low, and they suffered from many infectious diseases, which ultimately led to short lives. Under these circumstances, as nutrition officials and others actively worked to improve nutrition, the country was able to escape poverty through rapid economic growth, the food situation improved, and processed livestock foods from Europe and America were introduced, freeing the people from the state of low nutrition. As a result, infant mortality, infectious diseases, and even stroke mortality were reduced. In other words, I argued that Japan had created excellent Japanese food with balanced nutrition by its own efforts through a movement to improve nutrition as well as to escape from poverty through economic development (Table 8.1).

After the symposium, an American nutritionist who was listening to the symposium said, "I could understand that Japan succeeded in improving its nutrition and became a country of longevity because the government drew up an excellent nutrition policy, dietitians and nutritionists worked hard to teach it, and the people believed in it and put it into practice. But this kind of thing is impossible in the United States," she said.

"Why can't you do it in America?"

Table 8.1   Key points of nutrition improvement in Japan

1	In an era of frugal diets with a heavy emphasis on staple foods, the Japanese suffered from deficiencies in protein, essential fatty acids, and various vitamins and minerals, and many nutritional deficiencies (such as beriberi and night blindness) developed.
2	The mortality rate of infants was high due to low nutrition, and many people died of tuberculosis and other infectious diseases because they lacked resistance.
3	The combination of low nutrition and excessive salt intake led to high blood pressure, strokes, and even death from stomach cancer, and the Japanese had a short life expectancy.
4	An excellent nutrition policy based on nutrition science was established, nutrition improvement was implemented as a national movement, and nutrition guidance was provided at home, in the community, and at group meal facilities such as schools, hospitals, and companies.
5	As a result, a state of excess or deficiency in energy and nutrients was resolved by introducing a moderate amount of high-nutrient Western food into a low-nutrient diet, and a Japanese diet with excellent nutritional balance was enveloped.
6	The Japanese people overcame both under-nutrition caused by poverty and over-nutrition caused by high economic growth to form the world's longest-lived nation.

"Americans turn to the left when Washington tells them to turn to the right, and registered dietitians (RDs) are professionals who work in clinical organizations and aren't passionate about improving the nutrition of healthy people."

The nutritional improvement in Japan has been achieved by mixing the traditional Japanese diet and its low-nutrient density with the nutrient-rich Western diet, and by providing nutritional guidance and education by dietitians in schools, hospitals, workplaces, and even in the community, which has put the brakes on the Westernization of the diet and created a moderate mix. I believe that this can be of great help to the countries of Asia and Africa, which are suffering from the "double burden of malnutrition", a mixture of thinness and obesity caused by rapid economic development. In other words, the reason why the Japanese people today continue to maintain the world's longest life expectancy is not that their traditional diet was inherently superior, but that through their own efforts they have created a new Japanese diet that is superior in terms of nutritional improvement.

For example, during the post-war period of food shortages, imported food was used for school lunches in order to give priority to children who had a future, and in local communities, nutritionists and dietary life improvement promoters who worked as volunteers drove "kitchen cars," which were buses given to them by GHQ and converted from the back into kitchens. They traveled to every corner of Japan, teaching people how to cook and spreading the knowledge of nutrition including for tube feeding, nursing care food (p. 45, Photo 3.4). On the other hand, at a time when improvement of diet alone was insufficient, they developed and spread supplements such as vitamin A, B_1 or iron, and actively worked to improve the state of nutrition.

8.5.2   Improving Dietary Nutrition and Nutrition Policy and to Japan Nutrition

In all countries, the poor suffer from vitamin and mineral deficiencies, and vitamin A deficiency is particularly serious, causing many children to lose their eyesight. Our country solved night blindness and Beriberi by improving the diet and introducing the "fortified food system" at the same time. In the 1960s, post-war malnutrition was almost completely solved, and no other country has solved war-related malnutrition in such a short time and in such an equitable manner. The improvement of nutritional status resulted in healthy and excellent young people, who became excellent human resources for the rapid economic growth that followed and won many medals at the Olympics.

The excessive Westernization of the diet led to overeating, obesity and an increase in lifestyle-related diseases. As a preventive measure, the "Nutrition Improvement Act" was changed to the "Health Promotion Act", "Health Japan 21" and the "Specific Health Examination and Specific Health Guidance" were promoted, and the "Basic Act on Nutrition Education" was established to provide nutrition education through both group and individual education, thus putting the brakes on excessive Westernization.

These Japanese initiatives are collectively referred to as "Japan Nutrition" and should be a model for the world.

8.6 Everyone Was Impressed

8.6.1 Closing Ceremony of ICD 2008

The final day of the "ICDA 2008 Yokohama" was a great success with all venues fully booked. At the closing ceremony, Pr. Shigeru Yamamoto, Chair of the Executive Committee, thanked the participants for their participation, and Ms. Marsha Sharp, ICDA Canada Representative Director, explained the history of ICDA's development, organization, mission, and future activities. The ICD flag was handed to Sandra Capra to indicate that the next meeting would be held in Sydney in 2012.

Further remarks were made by Ron Moen, ICDA USA Representative Director, Carole Middleton, ICDA UK Representative Director, and Mary-Ann Sorensen, ICDA Denmark Representative Director, before I concluded with a final thank you to all those involved.

In fact, until 3 h before the closing ceremony, I was struggling to create a farewell text. I was working at my computer to write a text that would explain the significance of the theme of this conference, "Realization of a Peaceful and Healthy 21st Century from the Perspective of Diet and Nutrition," and to express my gratitude for the cooperation of the participants and others involved. Then I began to type the last sentence on my computer keyboard, "Please never forget the four days we spent here in Yokohama in 2008, working earnestly to solve the world's nutrition problems". As I typed the last sentence on my computer keyboard, I felt a sense of relief and memories of the past 30 years began to appear like a magic lantern, tears welled up in my eyes and I could not type any more. I thought that I would cry on the podium, so I went to the restroom, let out all my tears, and tried to calm down.

But the effort was in vain.

8.6.2 Crying Out Loud

When I went up to the podium and gave the final phrase of my closing speech, I cried out loud in front of the public for the first time in my life. I had been taught from childhood that a man should not cry in front of others, so this was a complete surprise to me. However, at this time, I felt a strange pleasure. I was able to forget my shame and feel good. I decided that if I could feel this way, I would cry as hard as I could, no matter what people thought. While crying, I could clearly see the faces of the participants who were crying with me. Applause broke out, and the sound gradually became louder and louder, enveloping the hall.

As soon as I stepped off the stage, Dr. Hosoya said to me, "This is the most moving international congress I have ever attended". I received warm words from many people. At the end, a video showing the scene during the congress was played, and the participants' serious studious and smiling faces enjoying the event were shown on the screen one after another, closing the congress with a touching finale. I had spent almost half my life preparing for this congress, all the participants were impressed by the congress.

8.6.3 International Congress Attraction and Hosting Contribution Award

The day after the congress, I was waiting for the train at the platform as usual, and without thinking I started humming. I hadn't hummed for many years. The song that came out was "Smile," which was played when we all did gymnastics during the break in the opening ceremony. I was finally able to put down the really, really long and heavy load.

On December 9, 2009, the Japan Intergovernmental Tourism Organization (JETRO) presented us with the "International Congress Attraction and Hosting Contribution Award" for the best management of an international congress held in 2008 (Photo 8.3).

Photo 8.3 International Congress Attraction and Hosting Contribution Award Japan National Tourism Organization December 9, 2009

8.7 Contribution to Asia

I believe that our country is the most successful country in the world in improving nutrition. This is also the country that has shown that it is possible to build a long-lived country. By hosting the ICD in 2008, we learned that we should not just boast about it, but we should teach people in other countries how to do it and make it a goal for all people to have a long life. This is exactly what human nutrition aims to do.

8.7.1 Nutrition in Developing Countries

Particularly in developing countries in Asia and Africa, where hunger and stunted growth of children due to low nutrition are still unresolved, some wealthy people are suffering from obesity and diabetes. I was once told the following story by a representative from Africa at an international conference.

"Children are dying of malnutrition because they cannot eat. Even if they could, their learning ability would not improve and they would not be able to develop excellent human resources. If human resources are not nurtured, industries will not develop and people will not become rich. If we cannot become rich, we cannot buy food, and malnutrition cannot be solved. We can't crawl out of this hell of a vicious circle." Many of the country's elites who complain of such misery are large, obese people.

Developed countries are providing economic aid, food aid, and agricultural technical assistance to Asian and African countries. However, it is difficult to solve the nutrition problem. Even if the economic and food problems are solved, economic development creates economic disparity in society, leaving hunger and malnutrition among the poor unresolved, and increasing obesity and non-communicable chronic diseases due to overnutrition among the wealthy, resulting in increased medical costs. In the process of national economic development, advances in food processing technology and the increase in imported foods have made high-calorie foods with high sugar and fat content available at low prices, and new forms of obesity are emerging among the poor.

As a matter of fact, there is a country whose economy had been exhausted by many wars and whose cities had been burnt to the ground by enemy bombing and had lost everything. That was our country, Japan.

Japan experienced the same severe poverty and food shortages as other countries in Asia and Africa. The conditions may have been even worse in Japan, which has no natural resources. However, in less than 20 years after the end of World War II, Japan solved the problem of low nutrition and achieved high economic growth through the efforts of its policy makers and people, and at one time, Japan was called "No. 1". Although there are many factors that contributed to this miracle, there is one thing that can be said without a doubt as a characteristic of Japan. It is that only Japan tackled the problem of nutrition as a national policy in the process of transition from developing to modern, and then becoming a stable and prosperous country. It created a law called the "Nutrition Improvement Act", established a system to improve nutrition, trained a large number of nutritionists and dietitians to be responsible for it, and placed them as professionals in every part of society. In many developing countries that suffer from malnutrition, there are no professional nutritionists, or if there are, their numbers are small, and specific nutrition policies are often not implemented. In other words, there is little concrete effort to improve the nutritional status of the people, which would guarantee the lives and health of the people who support the nation. In contrast, Japan has been able to produce many excellent human resources by improving the nutritional status of the people even before economic development started, and this has raised the labor productivity and provided a foundation of the nation's development.

8.7.2 International Contribution to the Training of Registered Dietitians

Since the time I was working at St. Marianna University Hospital, I had been training dietitians from the Philippines, Indonesia, Thailand, Cambodia, and Africa as part of a project of the Japan International Medical Technology Foundation (JIMTEF). When I joined the faculty of Kanagawa University of Human Services, I was still thinking about how to make international contributions. In 2015 (Heisei 27), Professor Shigeru Yamamoto, a senior professor at Jumonji University consulted with me. There was a proposal to establish a training program for registered dietitians at Hanoi Medical University in Vietnam, and he asked if I would be interested in working with him. Due to the long and fierce war with the United States, the food situation in Vietnam was poor, and the country was suffering from energy protein deficiency, iron deficiency anemia, short stature, and vitamin deficiency. Hospital meal service was inadequate and clinical nutrition management was not being implemented. There were nutritionists who specialized in nutrition, but no dietitians.

As in other developing countries, the food situation in Vietnam is improving due to economic growth, but the poor and rural areas are still under-nourished, while the rich and urban areas are over-nourished. The result was a "double burden of malnutrition", with thinness, shortness of stature and anemias caused by undernutrition, and obesity, diabetes and atherosclerosis caused by over-nutrition.

Hanoi Medical University, Vietnam National Institute of Nutrition, Kanagawa University of Human Services, Jumonji University, and the Japan Dietetic Association signed a five-partners agreement on the establishment of a dietitian program in Hanoi on March 23, 2013 (photo 8.4). Ajinomoto Co., Inc. opened an endowed chair at Hanoi Medical University and helped to create a base to prepare for the establishment of the new department, and the Japan Dietetic Association provided financial support because it was difficult to obtain public support at the preparation stage. In order to modernize hospital meal service, it was also necessary to develop food hygiene management, and Kao Corporation set up a scholarship scheme

<The five partners agreement>
Hanoi Medical University
Vietnam National Institute of Nutrition
Kanagawa   University   of Human Services
Jumonji University
The Japan Dietetic Association

< Cooperation>
Ajinomoto Co., Inc.
Kao Corporation
Kanagawa Prefecture
JICA

Photo 8.4 Five-partners agreement on training dietitians at Hanoi Medical University (2013)

Photo 8.5 Lecture at Hanoi Medical University

[maybe 'program'] for international students to study at the graduate school of Kanagawa University of Human Services after graduating from Hanoi Medical University. The Kanagawa Prefectural Government supported training in Japan for teachers from Hanoi Medical University, and Japan International Cooperation Agency (JICA) helped us to prepare manuals on clinical nutrition and hospital meals, which are necessary when working in hospitals after graduation. The lectures at Hanoi Medical University were given by Japanese lecturers in English and translated into Vietnamese by Vietnamese teachers (Photo 8.5).

In this project, we supported the education and training of professionals in a developing country through a so-called industry-government-academia collaboration, and we exported 100% of the Japanese method of education and training of dietitians to establish the profession of dietitian abroad. The curriculum and textbooks were translated, and about 30 teachers and other related personnel were dispatched to teach classes and provide practical training. Although the teachers felt stressed about writing the textbooks and teaching in English, they were so impressed by the enthusiasm, excellence, and thoughtfulness of the Vietnamese students that none of them complained. As I went through the classes, I felt less and less uncomfortable with the idea of teaching in English. At the end of the class, students said, "We want to give you a present to thank you, but we don't have money to buy it," and they all sang us a song instead. They gave me a pencil sketch of myself, which has become my treasure. No one slept or talked to one another in class, and the way they listened to the teacher's words with shining eyes reminded me of the origins of education for young people who carry the nation on their shoulders. The Japanese teacher in charge was given the position of "visiting professor" by Hanoi Medical University.

After graduation, some of them go on to graduate schools in Vietnam or abroad, and some of them work in hospitals or government. There is no doubt that they will grow up to be leaders in public nutrition, hospital and clinical nutrition management, healthy eating and diet therapy, school meal programs, and nutrition research in Vietnam. I believe that the seed sown in Vietnam will definitely sprout and bloom in the same way that nutrition in Japan has done.

8.8 Economic Development Alone Will Not Solve the Nutrition Problem

From January 15–20, 2017 (Heisei 29), I went to Cambodia for an inspection tour and to lecture as part of the Nutrition Japan Public Platform (NJPP) organized by the Japan International Cooperation Agency (JICA). Here, as in other parts of Asia, the rapid economic growth had made the rich over-supplied in terms of food, but the poor were still suffering from malnutrition. In the special economic zone being promoted in the capital city of Phnom Penh, a Japanese company had moved in, and many young Cambodians were working in the factory there. Many of these workers were poor young men who had come from the countryside. Due to poor nutrition, they lacked physical strength and were often absent without notice. They were told by local doctors that they had iron deficiency anemia, but they had neither the awareness nor the knowledge to improve their diet. They need to eat a diet rich in iron, protein, vitamin B_2, vitamin B_{12}, folic acid, vitamin C, etc., which are effective in preventing anemia, but they had few opportunities to maintain an adequate diet.

I heard an interesting story from a local manager. When the company first entered Cambodia, workers complained that they collapsed at work because the local president did not pray to God. He immediately built a household shrine. When the collapsed employees were taken to the hospital, they were diagnosed with iron deficiency anemia, and the company felt the need to improve their diet and provided the workers with money for food. However, the young women and men sent the money back home, and their deficient diet did not improve. I suggested that the company create a food service system in the workplace that would provide nutritious meals, even if it cost the company some money. I explained to the person in charge that improving the diet would improve the health of the workers, which in turn would benefit the company by increasing the labor productivity of the workers.

In fact, the improvement of the nutrition of factory workers was an important issue in Japan during the Meiji and Taisho periods, and there was a field called "labor nutrition". As a result, companies hired nutritionists, built cafeterias in their factories, improved the nutritional content of menus, and provided nutritional education to workers so that they could work effectively, and the costs were paid as part of the employees' welfare benefits. In other words, the Japanese people created a social system that allowed them to continue to eat nutritionally balanced meals, both in school lunches when they were children and in the staff cafeterias at work. Industrial products made on the basis of such excellent human resources in Japan have come to be highly evaluated by the world as "made in Japan" with few defects.

We hear from experts and leaders in developing countries that as their economies develop and the incomes of their people increase, their diets will become richer and their nutritional problems will be solved. In Phnom Penh, fashionable cafes and restaurants are increasing on the streets, and the content of meals is beginning to westernize. This phenomenon can be seen in any developing country, and the problems of hunger and low nutrition to some extent can be solved. However, economic development without a proactive nutrition policy will only end up with the nutrition problem shifting from low-nutrition to over-nutrition due to the Westernization of the diet. When over-nutrition becomes a problem, the number of non-communicable chronic diseases, known as lifestyle-related diseases, increases, and medical and even nursing care costs increase, causing greater social problems. In addition, due to economic disparity and unbalanced health awareness and knowledge, there will be under-nutrition among young people and the elderly, resulting in a variety of nutritional problems. Therefore, even if the economy, industry and culture develop, malnutrition will never be solved naturally. We believe that it is impossible to make the people of a country healthy and happy unless a specific "nutrition policy" is established, based on the country's unique food culture and food environment, and encompassing food policy, health policy, and economic policy. In other words, a comprehensive nutrition policy is necessary (Table 8.2).

8.8.1   Introduction in Nature

In 2016, we received an unbelievable interview. Nature International published an article about our activities and introduced us to the world (Photo 8.6).

Table 8.2 Suggestions for nutrition policy for developing countries

1	Progress in nutrition research and education
2	Development of nutrition administration and planning and implementation of nutrition policy
3	Expansion of nutrition and hygiene education to the entire population
4	National Nutrition Survey and formulation of Dietary Reference Intakes
5	Improvement of nutrition management and hygiene management for group meals such as school lunches, industrial lunches, and hospital diets
6	Establishment of a clinical nutrition management system for injured and sick patients
7	Appropriate production, manufacturing, distribution, sales, and consumption of food and food products
8	Institutionalization and production of food for special use for purposes of nutritional supplementation and disease risk reduction, production, dissemination, sales, and consumption
9	Implementation of nutrition and health labeling
10	Training and utilization of nutrition researchers and nutrition professionals

Photo 8.6 "Introduction article in Nature international edition Document" Searching for a long, healthy life, Spotlight on Food Science in Japan. *Nature* **534**, 12–13, 2016

Bibliography

Program Book, 15th International Congress of Dietetics, 2008

Searching for a long, healthy life, Spotlight on Food Science in Japan. Nature 534, 12–13, June 2016

Chapter 9
Cutting-Edge Science and Technology and Personalized Nutritional Advice

Abstract Society 5.0 is a human-centered society that can develop economically while solving the problems of modern society through a system that closely integrates cyber space (virtual space) and physical space (real space). Society 5.0 aims to create a society in which people and things are connected by the Internet of Things (IoT), and in which the necessary information is provided via robots and other means using artificial intelligence (AI) based on a variety of knowledge and information that exceeds human judgment.

Nutrition improvement can be practiced more reliably using cutting-edge technology. AI has the ability to analyze a large amount of scientific data, identify problems, and tell us how to solve them. In order to survive in Society 5, nutritionists and dietitians can use AI and robots to improve scientific accuracy, and at the same time, create new social value that can cause behavior change through relationships of trust between humans that machines cannot. It is necessary to produce. To do so, it is necessary to study the theory of psychology and master the skills of counseling.

There are several models of behavior change theory, especially Behavioral substitution method, Operant enhancement method, cognitive behavioral therapy, self-monitoring method, behavior changes stage model and nudge theory are effective.

Registered dietitians should acquire counseling technology which are listening closely, parroting, sometimes organize the story, respect for silence the improvement target is one or two feasible steps, always find the results become an attractive person, nutrition counseling that is attentive to patients' thoughts.

Keywords Society 5.0 · Science and Technology Basic Plan · Internet of things (IoT) · Artificial intelligence (AI) · Behavior change theory · Stimulus control method · Reaction disturbance and habit antagonism method · Behavioral substitution method · Operant enhancement method · Cognitive behavioral therapy · Self-monitoring method · Behavior changes stage model · Nudge theory · Caseworkers · Counseling technology

T. Nakamura, *Japan Nutrition*, https://doi.org/10.1007/978-981-16-6316-1_9

9.1　Society 5.0 and Nutrition

The Government of Japan established the "Science and Technology Basic Act" in 1995 (Heisei 7). The purpose of this law is to formulate a "Science and Technology Basic Plan" every four years in order to implement systematic and consistent science and technology policies from a long-term perspective. So far, the plan has been formulated for the first period (1996–2000) to through the fourth period (2011–2015), and science and technology policies have been promoted in accordance with these plans. The 5th Basic Plan was approved by the Cabinet on 22 January 2016, and science and technology policy is currently being pursued according to its contents. In this basic plan, Society 5.0 was proposed as the future vision of society.

Shouldn't there be a 6th basic plan now in 2021.

9.1.1　Working Toward an Ideal Society

"Society 5.0 is a human-centered society that can develop economically while solving the problems of modern society through a system that closely integrates cyber space (virtual space) and physical space (real space)." In other words, it is an attempt to make the whole of social life convenient and comfortable by accumulating a huge amount of information emanating from real space into virtual space, analyzing this big data with artificial intelligence (AI), and feeding back the results of the analysis to people living in drop real space in various ways.

If we call hunting society, Society 1.0, then the agricultural society is called Society 2.0, the industrial society is called Society 3.0, the information society is called Society 4.0, and the future human-centered society is called Society 5. Specifically, Society 5.0 aims to create a society in which people and things are connected by the Internet of Things (IoT), and in which the necessary information is provided via robots and other means using artificial intelligence (AI) based on a variety of knowledge and information that exceeds human judgment. The idea is to use cutting-edge science and technology to solve the problems that our country is currently facing, such as the declining birth rate, ageing population, economic stagnation and environmental collapse.

9.1.2　Development of AI

The other day, when I called for a taxi from my home, instead of the voice of the receptionist as before, I got a robot voice replying, *"Mr. Nakamura, If you would like a car delivered to your home, as usual, just hang up and please wait."* When I hung up the phone, the AI found the location of the nearest car, from the car navigation system and 10 min later, the car arrived as usual.

When he arrived, the driver said to me. "I've been dictated to by AI. eventually, they won't need the office, and when they go automated, they won't need us either."

However, according to the driver's story, the number of customers temporarily decreased because some people turned off the service when the voice from the robot was played. It seems that some customers look forward to the voice of the friendly receptionist. It's going to take some time for it to be accepted by all.

AI and robotics are also becoming more prevalent in the field of nutrition. There are already "cooking robots" with arms that extend from both sides of the kitchen to follow instructions and cook. There are also apps on the market that allow you to take a picture of the food you are eating with your smartphone, and the AI will authenticate the picture, check your intake of food and dishes and calculate your nutritional intake. Furthermore, a method of evaluating the nutritional status by comparing this nutritional intake with the "Dietary Reference Intakes" and the reference amounts for dietary therapy is also about to be put into practical use.

AI will select the appropriate nutritional intake from the vast amount of guidance comments that have already been registered and convey them to the subject via a communication device. Such methods are now being put into practice.

I was once invited to the laboratory of a venture company. In front of a refrigerator, you answer questions from the refrigerator about your age, physique, health condition, physical condition, tastes, etc. The content of the conversation is analyzed by the AI built into the refrigerator, and a high-performance refrigerator is being developed that will propose a menu suitable for you, taking into consideration the condition of the ingredients stored in the refrigerator. It can also tell you how to cook and even tell you how to buy ingredients that are in short supply. The development of convenient cooking utensils and the spread of pre-prepared foods will continue, and a society in which consumers can easily access the meals they want is already in sight through the networking of local convenience stores, supermarkets, drugstores, restaurants, and cafeterias. Although the accuracy, validity, and effectiveness of these technologies are not yet certain, scientific innovation will certainly affect the work of dietitians and nutritionists.

9.2   AI and Robots Are Not the Ones to Fight

How should nutrition and diet professionals respond to these changes in society?

I believe that this is the key to the survival of the nutrition profession. The following is a list of issues that need to be considered.

9.2.1   Utilizing AI

First, the experts who will certainly be important in an AI and robot society are researchers and developers engaged in the development of such cutting-edge technologies. Some nutritional scientists and dietitians are already actively involved in the development of the AI-based devices and systems mentioned above. AI researchers and developers are expected to join nutrition experts, and nutritionists and dietitians specializing in AI and robotics will appear.

Secondly, expectations are rising for dietitians and nutritionists who can use AI and robots to provide nutritional guidance so that nutrition and diet can be practiced in a way that is easy to understand and enjoyable for the target person or target group. In other words, nutrition improvement can be practiced more reliably using cutting-edge technology, and dietitian is a profession that can provide nutritional guidance with even better results than currently. In order to improve nutrition successfully, it is of primary importance to provide nutrition guidance based on scientific evidence. AI has the ability to analyze a large amount of scientific data, identify problems, and tell us how to solve them. However, just being taught how to improve nutrition does not mean that your nutritional status will improve. It is necessary that the subject actually makes a behavioral change in accordance with the solution and that this is carried out in a sustained manner, so that new eating habits are formed. New eating habits involve the ability to make appropriate choices of foods, dishes and menus, and to repeat healthy eating behaviors without the need for specific awareness or knowledge. Brushing your teeth every morning is something that you have learned to do as part of your health routine, not something that you do consciously each time you remember to brush your teeth.

9.2.2   Strengths and Weaknesses of AI

If the features of AI are used in several families to ensure improved nutrition, the task can be carried out efficiently and effectively. For example, AI is good at comprehensively analyzing problems and telling us solutions based on subject interviews, diet and nutrition surveys, clinical examinations etc. However, the actual behavioral change by the subject requires nutritional guidance based on a comprehensive assessment of the individual's outlook on life, lifestyle, family environment, learning ability, personality and past habits. Moreover, a relationship of trust is necessary between the subject and the instructor, and very few people can be expected to decide to change the lifestyle they have developed over the years because of advice given by AI, no matter how scientifically good it is.

AI and robots also have the problem of not being able to proactively find issues and problems on their own compared to humans. Furthermore, they cannot follow the flow of the conversation or read between the lines, and are inflexible and

Table 9.1 Characteristics of AI and robots

1	Doesn't find issues or problems on its own
2	Doesn't work unless tasks and conditions are set for it
3	Is a "problem-specific tool" that solves only specific problems
4	Can't follow the flow of a story or read between the lines
5	Can't elicit achieve emotional empathy from people
6	Experts who are stubborn and inflexible

stubborn. Moreover, many people cannot develop emotional sympathy with AI (Table 9.1).

In this way, knowing the problems and weaknesses of the latest technology, while taking advantage of them to further enhance their own expertise, is crucial for the survival of future professionals. I believe that surviving as a professional while maintaining a certain social status in today's ever-developing society requires not only the universal values of the profession and the professional ability of the individual, but also the courage and wisdom to evolve and adapt ourselves to changes in society. Charles Darwin, who advocated the theory of evolution, left behind the wise words, "In the world of living things, it is not the strongest that survives, nor the cleverest that survives. Those whose changes happen to be adapted to their environment survive".

Players of shogi (Japanese chess) have already stopped fighting against AI. This is because they have realized that humans can no longer win. However, by practicing with AI, young chess players have become stronger and able to beat their seniors, which made the competition between humans more interesting and made the chess world more exciting. When locomotives and cars were born, they were often made to compete with track athletes and horse-drawn carriages. Today, however, there are no runners competing with sports cars, and each competes separately to make society more enjoyable. In other words, the direction of nutrition in the future is to think of AI and robots as a tool to enhance our own expertise, and to enhance our expertise in Society 5.

9.3   Teaching Nutrition to Create Behavior Change

Whatever the society, a universal value of the nutrition profession is for people to feel that they can trust the dietitian or nutritionist and look forward to meeting and talking with them. The true purpose of teaching nutrition is not simply to share one's expertise in nutrition with people. Computers and AI have already taught us how to spread nutritional knowledge, and the streets are full of information, including false stories. Therefore, it can be said that the acquisition of knowledge and skills to cause behavioral change is an essential matter for dietitians and nutritionists.

9.3.1 Behavior Change Theory

Since there are several models of behavior change theory, we have organized them in terms of nutrition guidance.

1. **Stimulus control method**

    Since behavior is changed by stimuli, this is a method to increase good stimuli and decrease bad stimuli. For example, when we have to get exercise, we increase visual stimulation by putting our favorite walking shoes by the door, and when we have to eat less, we try to control stimulation by avoiding passing a noodle shop on the way home.

2. **Reaction disturbance and habit antagonism method**

    For strong temptations to negative behaviors, we teach coping strategies that prevent the behavior from occurring. For example, when a person wants to "eat anyway," we instruct him or her be patient for 5 min and then go outside for a walk or do something he or she likes (response interference method).

    If the patient is unable to do this, we can instruct the patient to eat low-calorie salads, seaweed, or non-sugar sweeteners (habit-antagonism method).

3. **Behavioral substitution method**

    This is a way to replace unhealthy behaviors with healthy behaviors. For example, if a person has a habit of drinking alcohol or bingeing on sweets to relieve stress, they can transform it into a new healthy behavior, such as taking a weekend trip or getting involved in sports.

4. **Operant enhancement method**

    This is a method of setting behavioral goals and providing positive stimulation. It is similar to the method of feeding animals and making them learn tricks. In the case of humans, for example, if a person loses 5 kg in 6 months, they set a goal to go out with friends, go shopping, or receive a present, and then carry out the goal.

5. **Cognitive behavioral therapy**

    This is a method of transformation that makes the person aware of inappropriate health behaviors and practices, and moves them to more rational thinking and behavior. For example, if an obese person has a habit of unwillingly eating sweet every day, he or she may say, "How weak-willed I am! I shouldn't have eaten that. It's pathetic. I just can't do it." For those who are hopeless and believe that they can't do it at all, following process is a way to change them so that they can think and act positively.

    ① Life without sweets would be boring.
    ② Is there a cake that is low energy and tastes good?
    ③ Yes, I'll make my own homemade cake with vegetables.
    ④ This is a great way to get fiber, vitamins, and minerals.

⑤ It's fun to make something yourself
⑥ It's fun to have a lot of variety, so I do it over and over again.

6. Self-monitoring method

This is a method in which the subject records his or her words, actions, thoughts, and moods on a memo sheet (worksheet) and evaluates them. It is important to judge the subject's motivation, knowledge, learning ability, environment, etc. comprehensively to determine which method is suitable.

7. Behavior changes stage model

This is a model derived from research on smoking cessation in the early 1980s. It suggests that people go through five stages to change their behavior: "indifference stage," "interest stage," "preparation stage," "implementation stage," and "maintenance stage". This method is a way of working through each stage, keeping track of achievements and problems, in order to get as far as possible at each stage.

8. Nudge theory

Nudge means to prod someone lightly with an elbow for attention or as a signal, and nudge theory is "a method of helping people to voluntarily make better choices for themselves". It is a strategy based on behavioral economics that provides "a form of presenting choices" in a casual way. For example, rather than prohibiting or restricting the purchase or consumption of unhealthy foods, it aims to create an environment in which people can easily choose the "right food behavior".

For example, in supermarkets, healthy foods are displayed at eye level, shopping carts are separate with a string in the middle to create a dedicated space for fruits and vegetables, small sizes of alcoholic and high-sugar beverages are easily available, and a variety of low-sodium foods are displayed. In other words, it is a way to make people change their eating behavior to the right one and make it a habit with little awareness through modest guidance and warning.

9.4 Nutritional Guidance That Is More in Line with People, Which AI and Robots Cannot Do

What kind of nutritional guidance by registered dietitians can never be done by AI or robots? How can we become the kind of dietitian who makes people enjoy consultations and willing to go see the dietitian again?

I think that the key to overcoming AI and robotics lies in this area. AI-based nutritional counseling software has already been developed, but the comments made by AI are patterned, so users get bored after a few consultations.

I had the same problem when I started teaching nutrition in the hospital. The first time or two, patients would come in and then not continue. What kind of nutritional guidance should I provide? At that time, however, there were no textbooks or reference books on nutrition counseling for individuals. The purpose of nutritional guidance was to teach the subject proper nutrition in an easy-to-understand manner, specifically, the correct selection of foods, how to prepare menus, and guidance on cooking techniques. However, the more nutritional guidance was given, the more it became clear that it was difficult to improve the dietary habits of the subjects and to change their eating habits simply by teaching them the information and techniques described in textbooks.

In particular, I realized that I was completely powerless to help patients with anorexia, which at that time was developing as a harmful effect of extreme dieting. The patients had more nutritional knowledge and a stronger will than others, and they could practice with certainty, but they were suffering from malnutrition. In other words, it was impossible to correct their behavior with knowledge alone, because their biased knowledge had amplified into beliefs, which resulted in abnormal eating behavior. Concerned about this, I consulted Pr. Michiko Fukazawa, who was also a pioneer for medical caseworkers at St. Marianna University Hospital. Caseworkers are professionals who provide consultation and assistance to people with mental, physical, and social problems, and they possess advanced counseling skills based on psychology. On her advice, I read a book on counseling and thought about applying these techniques to nutritional guidance. In the end, I realized that the important thing in nutritional guidance is to build a trusting relationship with the patient.

Knowing that the words of a reliable dietitian can have considerable power to change the lifestyle of patients. I have organized the points for building a reliable nutrition consultation into eight items based on a series of difficult consultation cases. I think that it is necessary for registered dietitians to acquire such skills in order to live in the society 5.0

9.4.1 Nutrition Counseling Skills That Registered Dietitians Should Acquire

① Listening closely

The first 10–15 min of a nutritional consultation should be devoted to listening to the patient. Listening to the patient is not just listening to what they have to say, but it is also listening hard and making it easy for them to talk by saying things like, "I see," or "Is that so?" There are patients who are satisfied with being listened to. However, no matter how much I listen, there are patients who do not want to speak and do not start to talk.

For example, there was a patient with anorexia who came to the Nutrition Consultation Room from Sendai. At first, he did not speak at all and was remarkably thin, so we admitted him to the hospital. We walked in silence along the corridor to the ward, and when we arrived at the ward, I casually asked him what he liked.

Then one word: "Tennis."

"Right, well, I'll see you at the tennis courts over there tomorrow morning"

At 7:30 in the morning he showed up at the tennis court. After a few rallies, we sat down on the bench and gradually began to talk. It is important to keep in mind that it is not easy to initiate an honest conversation with a person.

② **Parroting**

Parroting is a method of repeating the words spoken by the patient. For example, when a patient complains that he cannot stop to take sweets, you should return the parrot as follows.

"I see. You can't help but indulge in sweet things, can you. I sometimes do that, too"

If I repeat the same thing that the patient complained about, he will think that the registered dietitian understands his worry and suffering, and the distance between patient and registered dietitian will be reduced.

③ **Sometimes organize the story**

Patients talk one-sidedly about their thoughts and feelings, regardless of the context of the conversation. The order of the conversation and the relationship between the patient and others may become incoherent, so it is important to organize, order, and confirm the conversation from time to time before proceeding to the next conversation.

"So, this is what you are trying to say, Mr./Ms. ○○." And then, we move forward by organizing the conversation and confirming it. In this way, it is possible to delve into the actual situation of the person's dietary habits and find out the points that need to be improved.

④ **Respect for silence**

It is important to maintain silence from time to time during the dialogue. For example, when, a patient tells you something he or she has never told anyone before, respect the silence. Then say something like. :

"Yes, this is a difficult subject, so you can take your time and think before you talk about it."

Then, when so me story after the silence, put the words such as the following.

"It must have been hard for you, and I'm so glad you told me about it."

"You've made a difficult decision"

⑤ **The improvement target is one or two feasible steps**

From a professional point of view, many problems exist for patients. However, in general, people remember only one or two words that they hear from others. If we are told too many things, such as "follow this and that," we are left with the impression that dietary therapy is difficult, and we lose the goal of what exactly should be improved by next time. If patients can't do anything, they don't feel like going to the consultation room.

For example, if there are five items that need to be improved, I will make a plan to improve them one by one over the course of five consultations. I narrow down the items to be improved to one or two before the next visit, and at the end of the consultation, I speak clearly so as to leave a lasting impression.

"Remember, until next time, alcohol only twice a week."

In addition, it is effective to set improvement targets based on a consensus system.

"I'd like to set a goal of losing two kilos of weight by next month, would you like that?"

"No, I will be so busy with work and eating out this month that it's impossible for me to lose 2 kg."

"Okay, so you think you can do about a kilo first?"

"I think I can manage one kilo."

"So let's say we both agree to lose one kilogram."

Thus, if goals are set in a collegial manner, patients will be highly motivated.

⑥ **Always find the results**

The key to sustaining nutrition counseling is to find results of nutrition counseling that can be seen.

"Mr./Ms.○○, what's changed in a month?"

"Nothing, my blood pressure's not dropping. My blood sugar's not dropping."

"The weight has come down, hasn't it?"

"No, my weight hasn't changed."

"How are you feeling?"

"Nothing has changed."

"So let's measure your body fat, shall we? . . .We've seen a slight drop since the last."

"What? My body fat gone down?"

As you can see from this conversation, when you come back to the office, you should find at least one thing that has improved. When you find something, praise the effort and explain the significance of the result. This is because the biggest reason why patients cannot continue is that they cannot see any results.

⑦ **Become an attractive person**

The registered dietitian herself should strive to be an attractive person who is interested in sports, arts, and even culture on a regular basis. That person should be a professional who has a lot to talk about, who enjoys conversation, and who wants to hear a story again. Nutrition coaching is not selling an item of value; it is selling intangible words and attitudes. People are willing to pay for meaningful words from an attractive and trustworthy person, but they are not willing to pay for someone who is the opposite, and they will not want to see them again.

⑧ **Nutrition counseling that is attentive to patients' thoughts**

Nutritional guidance does not mean showing off one's own nutritional knowledge and skills in front of others. When I was young, I once observed a group of senior nutrition instructors. There were many types of nutrition instructors: those who talked about their knowledge in a one-sided manner, those who drew chemical symbols of nutrients with pride, those who talked about difficult topics using technical terms, and those who boasted about past cases.

What is important is that we should aim to be able to provide nutritional guidance that is based on scientific evidence and attentive to the subject's body and mind, so that a person can live a full and meaningful life for as long as possible, no matter what disease or disability they may have. In this way, nutritional guidance will become enjoyable and interesting for patients, and something they want to receive again and again. AI and robots are excellent sidekicks for such nutritional counseling, but they cannot play the leading role.

Chapter 10
A Sustainable Healthy Diet

Abstract The Sustainable Development Goals (SDGs) under the 17 goals, 169 targets and 232 indicators were set out by the United Nations in 2015. Sustainable development is defined as promoting sustainable, inclusive and equitable economic growth by providing opportunities for all. These are emphasized that in order for everyone to live a healthy and happy life on this planet without being left behind, all areas must be related, coordinated and harmonized in order to solve their respective challenges.

It is widely recognized that malnutrition has a negative impact on hunger, poverty, health, medicine and well-being, and nutrition affects a wide range of other areas, including education, labor economy, gender, discrimination, climate change and the environment. "Nutrition for Growth Summit: Fighting Hunger through Business and Science" held in London in 2013, and the "Tokyo Nutrition Summit 2021" in Tokyo.

In 2000, Dr. Paul Crutzen described the present as the "anthropocene". This is a new era in which humans are influencing the global environment, ecosystems, and climate, rather than an era in which humans were influenced by nature. And this new era is putting the earth in a critical situation which we are now facing. In 2019, the Lancet shocked the world with its report "Food in the Anthropocene: The EAT Lancet Commission on Healthy Eating through Sustainable Food Systems". The suggested changes in food intake include reducing the consumption of unhealthy foods, reducing as much as possible the consumption of meat, which has a high environmental impact. The WHO proposed 16 guiding principles for a "sustainable healthy diet", comprising "health aspects", "environmental impacts" and "social and cultural aspects".

Recently, due to the spread of a new coronavirus infection (COVID-19), people were forced to leave their homes as in an evacuation, and faced the same nutritional problems as during disasters. In recent years, many natural disasters such as earthquakes, typhoons, and torrential rains have occurred in Japan, and many people have been affected by these disasters and forced to live in shelters or temporary housing.

Disasters are caused by a combination of hazards and vulnerabilities, so it is necessary to establish a system during normal times to reduce community vulnerability from becoming apparent. The Japan Dietetic Association-Disaster Assistance

© The Author(s) 2022

T. Nakamura, *Japan Nutrition*, https://doi.org/10.1007/978-981-16-6316-1_10

Team (JDA-DAT) is a team of experts in the event of a large-scale natural disaster. JDA-DAT set up in the Great East Japan Earthquake occurred on March 11, 2011.

Keywords Sustainable Healthy Diet · Anthropocene · The EAT Lancet Commission on Healthy Eating through Sustainable Food Systems · Zero carbon emissions · The World Food Program (WFP) · Japan Dietetic Assistance-Disaster Assistance Team (JDA-DAT)

10.1 Nutrition Is the Foundation of Well-Being

Nutrition is related not only to the prevention and treatment of diseases, but also to the maintenance of life and health, as well as to a rich and cultured life. This has long been a topic of debate among nutrition experts. However, in the twenty-first century, with the globalization of political, economic, and environmental problems, nutrition has become an inevitable and fundamental theme in solving these problems, and the eradication of malnutrition has been found to be effective in alleviating many other problems.

A prominent example of this was the "Nutrition for Growth Summit: Fighting Hunger through Business and Science" held in London in June 2013, and the International Food Policy Research Institute's "2014 Global Nutrition Report: GNR" based on the commitments made at the conference.

The report describes several cases of interest. For example, in improving nutrition in the Democratic Republic of Congo, Mali, Nigeria and Togo, direct investments in nutrition policy resulted in an internal rate of return of +13%; in the 2015 Copenhagen Consensus involving 17 European countries, the median estimate for 17 countries was that the lowest benefit relative to the cost of addressing nutrition problems was reported to have a factor of 60. In other words, for every $1 of nutrition-related investment, there was $60 of benefit. The African Union Commission and the World Food Program (WFP) reported that in Malawi, undernutrition reduced GDP by 10.3% in the year 2012, while in Brazil, on the contrary, medical costs for obesity were estimated to be about 2% of health and medical service costs, in Europe 2–4%, and in the United States 5–20%.

In other words, undernutrition, as well as overnutrition, increases health and medical costs and has a negative impact on economic activity. In addition, a prospective cohort study of more than 3000 people in an urban area of Brazil, conducted over a period of more than 30 years, found that good nutrition in infancy was associated with increased school attendance and about 30% higher income. Good nutrition has long-term effects on school attendance and earnings.

This report, which was compiled with the aim of eradicating malnutrition in the world, states in its preface as its "philosophy" that good nutrition is the foundation of human life.

Against this backdrop, the United Nations Sustainable Development Summit was held at the United Nations Headquarters in New York in September 2015, with the

Fig. 10.1 Future challenges for the SDGs: Environment and nutrition

participation of more than 150 heads of government, and resulted in the adoption of the 17-point document Transforming Our World: 2030 Agenda for Sustainable Development Goals (SDGs). Under the 17 goals, 169 targets and 232 indicators were set out (Fig. 10.1).

Sustainable development is defined in the document as "promoting sustainable, inclusive and equitable economic growth by providing opportunities for all, eliminating inequalities and raising living standards, promoting equitable social development and inclusion, and facilitating the integrated and sustainable management of natural resources and ecosystems". The term "inclusion" means placing an issue within a larger scope, and the SDGs must be achieved by reconciling the three elements of economic growth, social inclusion and environmental protection. The term "social inclusion" is used here to describe the idea that each citizen, including those in vulnerable positions in society, should be protected from exclusion, friction, loneliness and isolation, and should be included and supported as a member of society.

Many countries are now beginning to address the SDGs.

Looking at the 17 goals set out in the SDGs individually, these are issues that have been discussed for a long time and are not particularly new. The SDGs are characterized by the fact that they present issues in multiple domains in a single diagram. In other words, the SDGs communicate that issues in one area affect other areas, and that comprehensive and integrated efforts are necessary to solve each issue. The SDGs emphasize that in order for everyone to live a healthy and happy life on this planet without being left behind, all areas must be related, coordinated and harmonized in order to solve their respective challenges.

10.2   SDGs and Nutrition

It is widely recognized that malnutrition has a negative impact on hunger, poverty, health, medicine and well-being. In particular, in today's discussion of the SDGs, nutrition affects a wide range of other areas, including education, labor economy, gender, discrimination, climate change and the environment, so that improved nutrition is considered essential to achieving the Sustainable Development Goals.

The goals related to nutrition in the SDGs were organized as follows.

10.2.1   Goal 1: Ending Poverty

Improving nutrition in any region can help reduce poverty by improving the labor force, increasing incomes and raising wages. On the other hand, poverty alleviation can also improve nutritional status by providing people with a minimum level of food.

10.2.2   Goal 2: Reducing Hunger to Zero

Eradication of hunger for any country is a fundamental life-sustaining issue and the most important nutritional problem. The causes of hunger include war, refugees, poverty, ignorance, religion and values. These are not issues that can be solved by nutritionists alone, but require the unstinting efforts of many people. The first thing that nutritionists should do is to investigate and analyze the specific hunger situation, and then to develop solutions based on this information. For example, in times of emergency, such as war or civil unrest, emergency assistance in the form of food and nutrition supplements is necessary, and in times of calm, there is the need to improve agricultural productivity, improve distribution mechanisms, and make effective use of limited food resources, as well as the constant implementation of group feeding and nutrition education. In particular, the promotion of sustainable agriculture will be necessary to put an end to hunger, secure a stable supply of food, and achieve improved nutritional status.

Hunger is a condition of high incidence of energy deficiency, so-called marasmus due to a significant shortage of food and food sources. In particular, nutritional deficiency during the growth and development of the fetus, infant, etc. causes health problems for the mother and fetus. It is important to investigate and analyze the specific features and causes of food shortages, and to develop food policies and nutrition education that are in line with the actual situation. It is also important to dispatch and train nutrition experts who can formulate improvement plans.

On the other hand, improved nutritional status will increase labor productivity and improve the quantity and quality of agricultural and industrial products. The improved nutritional status of women, in particular, will help reduce underweight newborns and improve breastfeeding, thereby saving children from hunger and malnutrition.

10.2.3 Goal 3: Health and Welfare for All

It is necessary to maintain and improve the health status of all people throughout their lives, regardless of race, sex, age, injury or disability. It is also important to ensure that people always have good nutritional status, whatever their circumstances. For example, 45% of under-five deaths worldwide are related to malnutrition, and stunting in children is associated with the development of non-communicable chronic diseases (lifestyle-related diseases) in later life and reduced labor productivity in adulthood.

The "Nutrition for the first 1000 days" movement, which is becoming increasingly popular around the world, demonstrates the importance of improving nutrition during the 1000 days including fetal life and the first two years. If a pregnant woman is undernourished, her fetus is exposed to low nutrition, which increases the risk of low birth weight, and at the same time, constitutes a precondition for overweight, which contributes to obesity and the development of non-communicable chronic diseases in adulthood. Thus, undernutrition in young girls can have lifelong health consequences for the next generation.

In addition, a reduction in overweight and obesity reduces the prevalence of non-communicable diseases, and under-nutrition is associated with the development of infectious diseases (diarrhoea, malaria, acute respiratory infections, tuberculosis and HIV/AIDS). These infections in turn are associated with the development of nutritional diseases and death.

Nutrition is indirectly related to the following goals.

10.2.4 Goal 4: Quality Education for All

Early childhood nutrition is related to educational development, and improved nutrition leads to improved school enrollment and achievement. If the level of education in a country improves, nutrition education will spread and progress, and the nutritional status will improve. In Japan, the school lunch system introduced in nursery schools, kindergartens, and schools is a part of comprehensive education on nutrition and diet, and has been very effective. This method is attracting attention from around the world as a method of achieving significant results.

10.2.5   Goal 5: Achieving Gender Equality

Improving the nutritional status of young and teenage girls can enhance their ability to learn in school and this can be empowering in the workplace and wider society, helping to improve the social status of women. In agriculture, improving the nutritional status of women can also improve their status.

10.2.6   Goal 8: Job Satisfaction and Economic Growth

Nutrition is related to labor, labor productivity, and even personal income.

10.2.7   Goal 11: Creating an Environment in Which People Can Continue to Live

It is a joy of life to be able to continue living in one's familiar community with a sense of well-being. In order to achieve this, it is necessary to create a society and environment in which people can enjoy locally produced food that changes with the seasons, eat together with their families and neighbors, and have a functioning community. In recent years, there have been calls for the establishment of comprehensive community care systems in response to the aging of society, and the important thing is the revitalization of communities, for which nutrition and diet are important issues.

In addition to these issues, nutrition is indirectly related to "Goal 13: Taking concrete measures against climate change," "Goal 14: Protecting the abundance of the oceans," and "Goal 15: Protecting the abundance of land as well".

As discussed above, nutrition is an important issue that influences each of the SDG goals and underpins the bottom line of sustainable development.

10.3   "Sustainable Healthy Diet"

In 2000, the Dutch scientist Dr. Paul Crutzen, who was awarded the Nobel Prize in Chemistry, described the present as the "anthropocene". This is a new era in which humans are influencing the global environment, ecosystems, and climate, rather than an era in which humans were influenced by nature. And this new era is putting the earth in a critical situation. There have been five mass extinctions of life on Earth in the past, which were caused by natural processes and events such as large-scale tectonic movements, volcanoes, freezing, and meteorites. However, the sixth crisis which we are now facing, the Anthropocene, is caused by human activities on the earth, and humans themselves are causing the mass extinction of life.

The use of fossil fuels, depletion of the ozone layer, deforestation, and desertification by humans over the past few decades has led to the extinction of 1000 species a year, and it is predicted that 1/4 of all living species will be extinct by the end of this century. In order to overcome this situation, discussions on how to reduce the burden on the environment and create a sustainable society are becoming more and more active.

In January of 2019, the medical journal The Lancet shocked the world with its report "Food in the Anthropocene: The EAT Lancet Commission on Healthy Eating through Sustainable Food Systems". The report set out a picture of a diet that would sustain the health and culture of each region in 2050, when approximately 10 billion humans would be able to eat without excluding anyone. It proposed a food system in which nutrition, diet, health, and the environment are in a win-win relationship. It weighed the "contribution to health" and the "burden on the global environment" that diet brings about, and proposed a system of nutrition and diet that balances the two.

The suggested changes in food intake include reducing the consumption of unhealthy foods like red meat and sugar, reducing as much as possible the consumption of meat, which has a high environmental impact, increasing consumption of fruits, vegetables and pulses, which have a high nutrient content, and a moderate intake of milk and dairy products (Table 10.1). The EAT committee also states that countries around the world will need to work to develop agricultural, cooking, distribution, processing, and menu options that will "reduce water use," "reduce nitrogen and phosphorus pollution," "produce zero carbon emissions," and "reduce methane and nitrous oxide emissions".

Based on these principles, the "Plenary Healthy Diet" (Healthy Diet for the Planet) was published. It recommends half of the plate being fruits, vegetables, and nuts, and the other half consisting primarily of whole grains, vegetable proteins (beans, lentils, legumes), unsaturated vegetable oils, moderate amounts of meat and dairy products, and additional sugars and starchy vegetables. This diet is similar to vegetarian and vegan diets, but states that personal preferences and local climate and culture should be respected.

The WHO proposed 16 guiding principles for a "sustainable healthy diet", comprising "health aspects", "environmental impacts" and "social and cultural aspects" (Table 10.2). The background to this proposal is as follows. The food system, which currently supports a world population of 7.5 billion, is a major cause of poor health and environmental degradation. The current food system makes diet-related non-communicable diseases: NCDs such as obesity, diabetes, and heart disease, the leading causes of death worldwide, while creating a situation of low nutrition for 800 million people, releasing 20–35% of global greenhouse gas (GHG) emissions, and accounting for about 40% of the earth's ice-free land area. And this system pollutes land, rivers and oceans with excessive fertilizers, making it the largest contributor to biodiversity loss. Without a transformation of the modern food system, sustainability and healthy eating are impossible, the report said.

Table 10.1 Reference values for a sustainable healthy diet (EAT Lancet Committee Proposal) (2500 kcal/day)

Food composition	g /day (acceptable range)	kcal/day
Whole grain		
Rice, wheat, corn, etc.	232 (0–60% energy)	811
Root vegetables, high starch vegetables		
Taro, cassava	50 (0–100)	39
Vegetables		
All vegetables	300 (200–600)	
Green and yellow vegetables	100	23
Red and orange vegetables	100	30
Other vegetables	100	25
Fruit	200 (100–300)	126
Dairy foods		
Milk and dairy products	250 (0–500)	153
Protein food		
Cattle· Sheep	7 (0–14)	15
Pork	7 (0–14)	15
Chicken and other poultry	29 (0–58)	62
Egg	13 (0–25)	19
Fish	28 (0–100)	40
Beans		
Dry beans	50 (0–100)	172
Soy products	25 (0–50)	112
Peanuts	25 (0–75)	142
Nuts	25	149
Additive fat		
Palm Oil	6.8 (0–6.8)	60
Unsaturated fats	40 (20–80)	354
Animal fat	0	0
Lard· Beef tallow	5 (0–5)	36
Additional sugar		
All sweets	31 (0–31)	120

As populations grow and become more urbanized, with greater consumer activity, the consumption of more high-impact foods, especially meat, fish, eggs, milk and dairy products, sugar and oil, will increase, placing a greater burden on the planet that produces them. Moreover, the shift to such a diet increases the risk of obesity and NCDs.

Ultimately, it is necessary to establish a relationship whereby a transition to a healthy diet reduces the environmental impact and a diet with a low environmental impact also becomes a healthy diet. For example, as a number of studies have pointed out, reducing meat intake can reduce greenhouse gas emissions and also reduce the risk of atherosclerosis because it reduces the intake of saturated fatty acids.

Table 10.2 WHO guidelines for a sustainable healthy diet (2019)

Health aspects	
1	Breastfeeding should be initiated shortly after birth and continued until the child is 6 months of age, with breastfeeding continuing to 2 years of age and beyond, combined with appropriate complementary feeding.
2	Balance throughout the food groups with a variety of non-processed or minimally processed foods, while limiting highly processed food and beverage products.
3	Including whole grains, legumes, nuts, and an abundance and variety of fruits and vegetables.
4	Including moderate amounts of eggs, milk and dairy products, fish, and red meat.
5	Safe and clean drinking water.
6	Adequate, but not excessive, amounts of energy and nutrients to meet the needs of an active and healthy life for growth and development and for overall lifestyle.
7	It is in line with WHO guidelines to reduce the risk of diet-related non-communicable chronic diseases and ensure the health and well-being of the general population (fat: up to 30–35% of total energy, transition from saturated to unsaturated fats, free sugars: less than 10% of energy, salt: less than 5 g).
8	Minimal levels of pathogens, toxins, and other substances that pose a risk of causing foodborne illness, or if possible their elimination.
Impact on the environment	
9	Greenhouse gases, water and land use, nitrogen and phosphorus use, and chemical pollutants within the target setting.
10	Protecting the biodiversity of crops, livestock, food of forest origin, and aquatic genetic resources. Avoiding overfish of fish and marine animals.
11	Minimizing the use of antibiotics and hormones in food production.
12	Minimizing the use of plastics and their derivatives in food packaging.
Social and cultural aspects	
13	Reducing food loss and waste.
14	Methods of sourcing, producing and consuming food structured in accordance with the values of the Earth's cultures, cooking practices, knowledge and consumption patterns.
15	Easy to access and includes what you want to eat.
16	Ensuring that the allocation of time for the purchase and preparation of food and water, and the acquisition of fuel, is not influenced by gender issues.

10.4   Emergency Nutrition Management

In recent years, many natural disasters such as earthquakes, typhoons, and torrential rains have occurred in Japan, and many people have been affected by these disasters and forced to live in shelters or temporary housing. Recently, due to the spread of a new coronavirus infection (COVID-19), people were forced to leave their homes as in an evacuation, and faced the same nutritional problems as during disasters.

Disasters are caused by a combination of hazards and vulnerabilities, so it is necessary to establish a system during normal times to reduce community vulnerability from becoming apparent. In other words, it is necessary to establish a system to protect people who may require assistance in times of disaster.

The acronym "CHECTP" stands for the following: C:child, H:handicapped, E: elderly people, C:chronically ill, T:tourist (a person who does not understand the language), and P: pregnant. It is important to consider the nutrition and diet of these people in order to maintain a healthy life in the disaster area and to avoid disaster-related deaths (indirect deaths).

(1) Issues and measures for nutrition and diet in times of disaster

When a disaster occurs, food and drink are sent to the affected areas from various regions and countries as relief materials, and various kinds of food and dietary support are provided, and this is reported as a sign of recovery. However, this does not mean that the problems of nutrition and diet can be solved, because the supply of food alone does not guarantee the good health and nutritional status of the victims. For example, if food delivered to evacuation shelters is equally distributed to people of different sexes, ages, sizes, and health conditions according to the principle of equality, meals that meet individual nutritional needs will not be provided, and the response to those in critical need will be incomplete. The health and nutritional status of people with chronic diseases and disabilities deteriorate with prolonged evacuation.

Nutrition and dietary precautions in times of disaster can be outlined.

① Adequate hydration

Generally, evacuation shelters are opened in school gymnasiums or local community halls, and facilities are set up for temporary toilets, but the toilets are often far away, unclean, and unsanitary, and users try to use the toilet less often. Because of the psychological burden of incontinence, many elderly people limit their fluid intake, and many people suffer from fluid deficiency. As a result, they are more likely to develop fever and dehydration, and the combination of aging and stress increases the risk of fatigue, constipation, hypothermia, cardiovascular disease, and deep vein thrombosis/pulmonary embolism (economy class syndrome). It is necessary to educate and instruct people to actively replenish fluids through eating meals and drinking fluids.

② Guidance on nutritionally balanced meals

In general, the amount of food distributed in disaster areas is limited, and some foods can't be eaten because that some food is not delivered in time and people are not familiar with it, resulting in a decrease in overall dietary intake and a tendency to suffer from malnutrition. In terms of nutrients, the intake of protein, fat, vitamins, minerals, and dietary fiber will be insufficient, while the intake of carbohydrates, sugar, and salt will increase. In terms of foods, the intake of processed foods such as sweets and pastries, noodles, instant noodles, retort pouches, and canned foods will increase, while the intake of meat, seafood, milk, dairy products, vegetables, and fruits will decrease.

In addition, elderly people and patients with chronic diseases such as dysphagia, diabetes, kidney disease, hypertension, and dyslipidemia require dietary treatment, but this is difficult to implement in disaster areas.

In order to solve the problem, proper stockpiling and distribution of food at evacuation shelters and disaster areas, as well as the management and operation of school lunches, will be important. In particular, it will be necessary to properly procure, manage, and deliver support supplies such as special[-] purpose foods, thickening agents, and thickened liquid diets to those who require special treatment. In addition, it is important to be able to respond to individual requests by using pack cooking (a vacuum cooking method in which food ingredients and seasoning solution are placed in a high-density polyethylene bag and cooked easily using an electric pot, etc.).

③ Strict adherence to food hygiene

From the viewpoint of preventing food poisoning, hygiene management is important, and it will be necessary to cooperate with administrative agencies in the affected areas to take action. Since temporary kitchens will be set up in local assembly halls and school gymnasiums, and hygiene control will be difficult, sufficient consideration should be given to food hygiene when purchasing, storing, cooking, distributing, and serving food, and washing dishes and cooking equipment.

④ Nutrition and diet to protect against infection

In COVID-19, due to the government's declaration of a state of emergency, leaving one's home was prohibited and socio-economic activities were suspended, and school lunch kitchens, restaurants, cafeterias, and shops reduced service or closed. As a result, food production and distribution declined, consumers bought in bulk, ate unevenly, and the eating of processed foods increased, resulting in unbalanced nutritional intake, weight loss or gain, and malnutrition such as nutrient deficiency or excess.

In order to prevent the onset and aggravation of infectious diseases, it is important to prevent infection by bacteria and viruses, and at the same time, it is necessary to enhance the resistance, or self-defense capability, of the body. The human defense function is influenced by various and complicated activities of the immune system, and different factors, such as malnutrition, extreme stress, fatigue, lack of sleep, lack of exercise, drinking and smoking, and illness, are related to the decline of the function.

(2) The birth and role of the Japan Dietetic Assistance-Disaster Assistance Team (JDA-DAT)

The Japan Dietetic Association-Disaster Assistance Team (JDA-DAT) is a team of experts who, in the event of a large-scale natural disaster (earthquake, typhoon, etc.) in Japan or overseas, quickly provide nutrition and dietary support activities in cooperation with medical, welfare, and governmental nutrition organizations at evacuation shelters, facilities, homes, temporary housing, etc. in the disaster area. When the Great East Japan Earthquake occurred on March 11, 2011, the Japan Dietetic Association set up a task force on March 15, and sent a group of three to four dietitians and nutritionists to Kesennuma City, Ishinomaki City, and Tono City, respectively.

Table 10.3 Improvement in nutrient intake with the support of dietitians and nutritionists

	1st survey (April 1–12)	2nd survey (May 1–20)
Energy	1546 kcal	1842 kcal
Protein	44.9 g	57.1 g
Vitamin B$_1$	0.72 mg	0.87 mg
Vitamin B$_2$	0.82 mg	0.96 mg
Vitamin C	32 mg	48.4 mg

In order to assess the effect of the dispatch of dietitians, we calculated the nutritional intake in evacuation shelters in Miyagi Prefecture. In the first survey from April 1 to April 12, when the support system was inadequate, the average daily intake was 1546 kcal of energy, 44.9 g protein, 0.72 mg vitamin B$_1$, 0.82 mg vitamin B$_2$, and 32 mg vitamin C, whereas in the second survey from May 1 to May 20, when JDA-DAT support was advanced, there was an improvement in energy, protein and many micronutrients (Table 10.3).

In municipalities in Iwate and Miyagi prefectures, rice and miso soup were supplied to evacuation shelters by the Self-Defense Forces. As a result, although energy and carbohydrates were sufficient, there was a marked lack of protein, vitamins, and minerals. The Japan Dietetic Association persuaded the Self-Defense Forces to introduce vitamin-fortified rice to be used as white rice for cooking, and at the same time distributed packs of nutritional supplements for home use. As a result, we were able to distribute vitamin-fortified rice to about 2.7 million people and supplement packs to about 1.34 million people.

JDA-DAT is currently responsible for the following roles.

① Coordination of food aid

In general, food aid to the affected areas is provided without assessing the local needs and nutritional status, so it focuses on carbohydrate foods such as rice, instant noodles, and sweets, and lacks protein foods, vegetables, and fruits. Thus the food supplied is often inadequate and there is a lack of coordination among evacuation shelters.

② Correcting food distribution based on strict equality

Rather than distributing all relief supplies equally, we will distribute food according to the individual circumstances of evacuees, considering their age, nutritional and health status, and eating capacity.

③ Distribution within the best-before date

In order to solve the problems of ①~③, it is necessary to comprehensively understand the delivery, storing, and consumption of foods in evacuation shelters, determine the excess or shortage of various foods, and adjust the food distribution so that it is nutritionally balanced. JDA-DAT will improve the nutritional status of the affected people by coordinating with the government, other organizations, and evacuation shelters to provide comprehensive nutrition and food management.

④ Dealing with high-risk individuals

Some of the affected people are at high risk for nutritional and dietary problems such as dizziness, fatigue, oral ulcers, colds, loss of appetite, nausea, constipation, diarrhea, anemia, bedsores, emaciation, aspiration, chewing problems, dehydration, edema, and tube feeding. JDA-DAT will conduct a detailed nutritional assessment of such subjects and adjust the foods to be provided as well as utilizing special foods as necessary. JDA-DAT receives assistance from the supporting members of the Japan Dietetic Association in the form of supplements and special foods for the sick.

⑤ Support for physical and mental well-being

The key to nutrition and dietary support during emergencies is to be attentive to the physical and mental needs of survivors. JDA-DAT is committed to providing humanitarian, supportive, and practical assistance to people under severe stress, based on Psychological First Aid (PFA).

⑥ Training and Collaboration

In an emergency, training sessions are held to train JDA-DAT leaders from among the JDA-DAT staff of each prefectural association of dietitians so that they can promptly conduct support and relief activities (as of June 30, 2020, 3303 people have been trained; Photo 10.1). In addition, JDA-DAT has deployed seven

Photo 10.1 Training session to develop JDA-DAT leaders

Photo 10.2 Disaster relief vehicle equipped with a kitchen unit

emergency disaster relief vehicles (Photo 10.2), which were donated by individuals and companies. These vehicles will be used during normal times and as needed in emergencies in order to provide mobility. In addition, an "Agreement on Nutrition and Dietary Life Support Activities in Times of Disaster" has been concluded with each local government and prefectural association of dietitians to ensure prompt cooperation in times of disaster. As of June 30, 2020, we have concluded agreements with 15 prefectures and 3 cities.

(3) Activities of JDA-DAT

When an emergency situation occurs, JDA-DAT will go to the affected area within 72 hours after the disaster and work under the direction of the administrative dietitian in the affected area, after receiving a request from the administration and dietitian associations in the affected area and cooperation requests from the Ministry of Health, Labor and Welfare. The main activities are to accompany medical rescue teams to evacuation shelters, provide nutritional counseling, conduct hygiene management, transport relief supplies, set up special nutritional food stations, and, at the request of the Ministry of Defense, provide support to disaster victims using the Hakuo (a hotel-ship), and facilitate collaboration with other societies, NGOs and companies.

Under the JDA-DAT, a "special nutritional food station" has been established to procure, deliver, and distribute special nutritional foods required by people with special needs, such as breast milk substitutes (powdered milk, liquid milk), baby food, low-protein food, allergen-free food, thick liquid food (including for tube feeding, nursing care food, food for people with swallowing difficulties, and food to be used with thickeners). For example, if there are elderly people in evacuation shelters who are unable to eat, they can be offered a sample of the soft food from a retort pouch. After they taste the food, the possibility of eating, chewing, swallowing, preference, etc. should be confirmed and the food should be provided. In other words, we provide food and guidance according to the needs of the individual.

Bibliography

2014 Global Nutrition Report (GNR) (2014) International Food Policy Research Institute

2018 Global Nutrition Report (2018) International Food Policy Research Institute

Katsuhiko S (2018) Nutrition and biodefense. In: Toru S, Suzuki K (ed) "Infection and Biodefense", a course in nutrition management, Kenpakusha, pp 73–80

Nakamura T (2011a) Launching the emergency task force for the Tohoku Earthquake. J Jpn Det Assoc 54(4):1

Nakamura T (2011b) Nutrition and dietary management in times of disaster. Vitamin 85 (9):459–462

National Institute of Health and Nutrition, Japan Dietetic Association, 2011 "Manual on Nutrition and Dietary Support in Times of Disaster", Flow of Activities for Meals and Nutritional Support in Times of Disaster, "Nutrition and Dietary Habits to Prevent Health Problems Arising in Evacuation: Leaflet for the Elderly" https://www.dietitian.or.jp/assets/data/learn/marterial/h23evacuation5.pdf

Office of Lifestyle-related Disease Control, General Affairs Division, Health Bureau, Ministry of Health, Labour and Welfare, 2011 "On the immediate target nutritional reference quantities for planning and evaluating the provision of meals in shelters" https://www.mhlw.go.jp/stf/houdou/2r9852000001a159-img/2r9852000001a29m.pdf

PFA: U.S. National Child Traumatic Stress Network, U.S. National Center for PTSD, Psychological First Aid Implementation Guide, 2nd Edition, translated by Hyogo Kokoro no Care Center, March 2009. Available at: http://www.j-hits.org/ (Original):National Child Traumatic Stress Network and National Center for PTSD, Psychological First Aid: Field Operations Guide, 2nd Edition. July, 2006. Available on: www.nctsn.org and www.ncptsd.va.gov

Willet W et al (2019) Food in the Anthropocene: the EAT-Lancet Commission on healthy diets from sustainable food systems, Lancet, Published online January, 16. https://doi.org/10.1016/S0140-6736(18)31788-4)

World Health Organization (2019) Food and Agriculture Organization of the United Nation, WHO, "Sustainable Healthy Diets Guiding: Principles"

Chapter 11
For Those Who Hope to Study Health, Medical Care and Welfare

Abstract I wrote a message for young people who are aiming to become specialists in health, medical care, and welfare.

1. Dreams are not something you can see or achieve, but something you talk about
2. Develop human resources with intelligence
3. Why do we keep learning?
4. Challenges from the unknown
5. Chance and failure create great discoveries
6. Helping one another to save lives
7. Maintenance and promotion of physical and mental functionality and nutrition
8. There may be an energetic element in food

Keywords Results-oriented approach · A world without nuclear weapons · Nobel Peace Prize · Humanity · Intelligence · Intellect · Intellectual revolution · Captain Cook · Dr. Jame Lind · Darwin · The Galapagos Islands · Jean André Dumas · Albumen · Kizuna · View of impermanence · Homo sapiens · Living together · Starvation experiment

11.1 Dreams Are Not Something You Can See or Achieve, But Something You Talk About

I believe that dreams are not just something to be seen, nor can they always be achieved, but rather they are meant to be "talk".

Our ancestors, in the midst of poverty or social instability and absurdity, 'had many dreams' and told them to many people. When reality became too severe, they had no choice but to express their dreams as stories. In recent years, however, young people have stopped talking about their dreams, and adults in particular have come to disregard dreams as fantasy and unrealistic stories. I believe that the background to this is the "results-oriented approach" that began at the end of the twentieth century. The emphasis is on planning and strategizing in such a way that final results can be generated rather than making unfounded, dream-like plans.

© The Author(s) 2022

T. Nakamura, *Japan Nutrition*, https://doi.org/10.1007/978-981-16-6316-1_11

The question that comes back to me when I tell my dream story is, "What do you measure outcomes by?"

The methodology of planning is to set feasible numerical targets and to work towards them. This method has been implemented in many areas because it allows us to define specific goals, to evaluate them easily, and to achieve results in a short period of time. In recent years, however, problems have emerged. This method alone does not produce long-term results, and it is difficult to draw a picture of what kind of world will be born in the end. This may be the reason why modern society is unstable, lacks an end goal, and is becoming chaotic.

I feel that times are changing and the time has come for us to "talk" about our dreams once again. It was on April 5, 2009, when I heard then U.S. President Barack Obama speak to a large crowd in Prague about his dream of "a world without nuclear weapons". He was awarded the Nobel Peace Prize, despite the fact that he was merely talking about a dream that is not feasible or achievable today. It was the first time in the history of the Nobel Prize that a Nobel Prize had been awarded for a dream with no apparent outcome.

In fact, the basis of his speech was an opinion piece published in the Wall Street Journal two years earlier by four leading American intellectuals, who argued that "we should aim for a world without nuclear weapons". They were experts who had long been engaged in U.S. defense policy, and at first, they were hesitant to publish such an unrealistic paper. However, they thought that the realization of a society without nuclear weapons should be presented as the path that human beings should take, even if it was an unrealistic story.

No matter how sick or disabled a person may be, or how old they be, or whether they have lost the use of their limbs or have lost their memory, as long as they have the will to live and the remaining functions, they can live a happy and vibrant life without being discriminated against. It may be a dream to create such a society. However, I believe that if we all talk about this dream and make constant efforts, it will come true one day. Even if we can't achieve it, I believe that if we talk about our ideal dreams publicly, people will sympathize with us, turn in the same direction, and become the driving force of a society that moves toward our ideals.

Students who have entered a vocational school or university and are about to become professionals are in the transition period of realizing not the dreams of children, but those of adults. There is plenty of time in the coming years of student life. In addition to your family, there are many people on campus who are willing to listen to your dreams, including your instructors, classmates, seniors and juniors, and people in the community. If you talk about your dream, you will give yourself the task of making it come true, and you will be able to take it seriously. Furthermore, the more you talk about your dream, even if it is a small one, the more people will celebrate with you when it comes true, the more people will grieve with you when it falls apart, and the more true friends you will make who will share your dream with you. And your life will become richer.

11.2 Develop Human Resources with Intelligence

Humanity is facing some of the most difficult problems it has ever experienced, including declining birthrates, aging populations, the collapse of local environments, economic disparity, international tensions, and unknown infectious diseases. All of these issues are unavoidable for young people who aim to become professionals in the fields of health, medicine, and welfare. Moreover, each of these problems is subtly related to problems in other fields, and many of these problems are characterized by the fact that no single solution can be found. In other words, in the future, we will have to learn from past problems that have been solved, while at the same time tackling complex and diverse problems for which there is no correct answer.

Already, many experts and scholars are tackling the intractable problems posed by modern society. For example, research is underway to use cutting-edge technologies in biotechnology and robotics to change organisms themselves, and regenerative medicine to repair and reinforce malfunctioning parts. Attempts to find answers through the analysis of big data and the use of artificial intelligence in all areas of our lives to explore more efficient and rational ways of living are already being put to practical use. In 2000, Eduardo Kac, a bioartist from Brazil, had the idea of creating a fluorescent-colored rabbit as a work of art. He commissioned a French genetic engineer to do the job, and the scientists took an ordinary rabbit embryo and implanted a green fluorescent jellyfish gene into its DNA. The result was a glittering rabbit that emitted a brilliant green fluorescent color. It is only a matter of time before such technology is applied to humans, and this technology can be used not only for art but also for the prevention and treatment of diseases. However, we should not forget that such technology has the danger of creating a person who suits our preferences.

In other words, the knowledge and technology that humans have acquired through science and technology have infinite possibilities, but they also have infinite dangers. This is the kind of age we live in today.

I read a book titled "Improve Your Intellect" written by Hiroshi Tasaka. He says that it is not "intelligence" that is necessary for a doctor, but "intellect". "Intelligence" is the ability to answer questions that have answers quickly and correctly, while "intellect" is the ability to keep asking questions that have no answers. He says that the conditions for improving intellect are thought, vision, aspiration, strategy, tactics, technology, and finally, "humanity".

Intelligence can be contained in a computer, and eventually artificial intelligence will manage it. However, intellect can only be attained by learning it through one's own efforts. I think that the qualities that modern society needs in human resources are intellect that can mobilize scientific knowledge and technology, enhance innovation, and actively tackle difficult problems. Come to think of it, there is an old saying,

"That person has an intellect and is an honorable person," which is used to express high esteem, but saying someone is an 'intelligent person' is not an ethical evaluation.

11.3   Why Do We Keep Learning?

People always dream for the future. Some of them are short-term plans and some are long-term dreams. In order to realize these dreams, various efforts are necessary, and the central part of these efforts is to "learn" various matters. However, it is not so long ago that people began to think systematically about the significance and methods of "learning". In the past, many people thought that it was more important to "believe" than to "learn". They believed in the teachings of God, Buddha, sages, [and] rulers, and prayed for their dreams to come true. This custom can still be seen at shrines on New Year's Day and before examinations. About 500 years ago, what is known as the "intellectual revolution" began in Europe. It was the scientific method of observing the phenomena of the world, searching for the factors and causes, discovering the laws, and using them to advance our own lives. Through this, human beings came to know the significance and value of "learning". By learning, they developed science, caused the industrial revolution in Europe, developed modern civilization, and became able to live a rational and affluent life. The reason for such an "Intellectual Revolution" in Europe lay in the preceding Age of Discovery. In those days, when European conquerors set out on a voyage, they brought aboard scholars of astronomy, geography, meteorology, botany, and anthropology, as well as sailors. In other words, conquerors not only wanted to conquer the unknown world and obtain gold, silver and treasures, but also wanted to learn new things.

The effects were first seen in the sailors themselves. Half of the seafarers died during a voyage from incurable diseases where the soft tissues of the body began to bleed, teeth fell out, wounds opened up, jaundice set in, and limbs became ineffective. Between the sixteenth and eighteenth centuries, some two million sailors died of this disease. At this time, Captain Cook of the British, following the advice of Dr. Jame Lind, gave the sailors citrus fruits, a local folk remedy of the time. As a result, the disease was completely prevented and cured. As we now understand, the incurable disease was scurvy caused by a deficiency of vitamin C due to the lack of fresh fruits and vegetables during the voyage. By learning how to prevent scurvy, the British were able to sail vast distances and acquired a vast amount of knowledge and data from all over the world, becoming the leaders in modern science. Similarly in physics, Newton discovered the law of the conservation of energy, and this idea led to the discovery of life energy, which became the starting point for nutrition. In biology, Darwin himself was on a voyage to South America and the Galapagos Islands, where he made observations that led to his theory. The theory of evolution made it clear that humans were not created by God, but evolved from primates that had adapted to their environment.

The origin of the fact that present society has become richer and more comfortable than before is not only because we "believe", but also because we have started to "learn", and I believe that therein lies the meaning of our continuing to learn until we die.

11.4 Challenges from the Unknown

The education and training of professionals engaged in health, medical care, and welfare have been carried out mainly in technical schools. Doctors were also trained at what are called 'medical colleges'. In recent years, however, nurses, dietitians, physical therapists, occupational therapists, social workers, and others have come to seek university education, just as doctors have shifted to education at university medical schools. The progress and systematization of each discipline, which is the foundation of professional education, have led to a shift to four-year universities, as two or three years are not sufficient, and recently there has also been a call for more graduate education.

There are three main reasons for this.

The first reason is that knowledge and skills in each specialized field have become more advanced, and an educational period of two or three years is insufficient, and even four years of undergraduate education tend to be insufficient, so the need for education at graduate school has emerged.

The second reason is that at universities, students can gain knowledge and skills that are related to specialized fields and support their expertise, such as philosophy, ethics, life science, constitutional law, sociology, statistics, and languages, so-called liberal arts courses. Mastering these basic studies is essential for expanding and deepening one's field of specialization, and is necessary in today's society, where there are calls for cooperation with other fields.

The third reason is that it is essential to become aware of new methods and technology while also learning that there are things that humans do not yet know. In other words, it is necessary to study the unknown, to advance the field, and to learn that this advancement can make people healthier and happier. For 132 days from September 19, 1870, the Prussian army attacked the citizens of Paris who had risen up against the Prussian army, and Paris was besieged.

The citizens ate up all the food they had stored in their homes and began to eat cats and rats, and there was even a "cat and rat cookbook". They also went to the zoo and ate animals such as horses, elephants and lions. Inside the barricades of the besieged city, there was Jean André Dumas, who was a famous nutritionist. Seeing the people around him dying, he decided to save infants alone, and created the world's first artificial milk. It was called "Albumen," and it was made by emulsifying various proteins with oil and sweetening them with sugar.

But even after being fed this milk, the infants were dying in droves. The current knowledge of nutrition could not save the children's lives. The reason for this was that nutrition at that time only knew about carbohydrates, fats, and proteins as energy sources. Nutritionists then thought that people could live as long as they consumed energy sources. This was a tragedy caused by immature nutritional science. However, because of Dumas's seemingly reckless challenge, people learned that there were nutrients that they did not yet know about, which led to the subsequent discovery of vitamins and minerals.

What is necessary for educators and researchers at universities is to have the humility to believe that there are unknowns in academia, the courage to try to learn about them, and the interest to continue investigating.

11.5   Chance and Failure Create Great Discoveries

Lately, I've been worrying about something.

That is, thanks to the advancement of IT technology, if you enter a keyword on your computer or smartphone, the machine will search for it and give you the answer instantly. If you become accustomed to this way of accessing information, you may mistakenly believe that every problem or issue has a correct answer and that you can easily get it if you just enter the keyword.

Before the development of IT, if you did not understand something, you had to ask the opinion of a teacher or an expert, spend time in the library, read all the relevant literature and books, and think about it. We had to think about this and that to get the answer. Scientists spent a lot of time, made repeated mistakes, and twisted and turned to get the answer. The answer may have been incomplete, inadequate, or incorrect. However, because I had spent so much time and effort to obtain the answer, I was satisfied with it and it was persuasive to others. In addition, in the process of deriving answers, I sometimes made unexpected discoveries by searching wrongly or looking wrongly. Recently, our country has been producing a Nobel laureate every year. The story that often comes out from the laureates is that if you keep challenging yourself, you will make great discoveries, but the process requires chance and failure. Trying to seek answers by the usual means that most people use will only yield the usual answers. It is through these seemingly futile challenges that we encounter chance and failure, and it is in the persistent search for the cause that new discoveries are made.

When you are a student, the more you learn, the more things you don't understand, and the more you stop, get confused, and worry. However, we must not forget that this is what will help us grow as human beings and professionals. I believe that the reason why human beings have evolved remarkably as intelligent animals compared to other animals is not only because we have evolved by adapting to our biological environment, but also because we have a strong will and the courage to learn about the unknown and to try to lead a better life, as well as the ability to take on new challenges.

11.6   Helping One Another to Save Lives

The twenty-first century began with the end of high economic growth, serious environmental problems, and an era in which humanity was searching for a new set of values and a new social framework. Then, on March 11, 2011, the Great East

Japan Earthquake struck. The science that had built our affluent society could not have foreseen the disaster, and nuclear power, which had brought together the best of science and technology, instantly became a high-risk energy source. However, in the midst of this darkness, we have found a ray of light in "kizuna". The word "kizuna": bond comes from the Chinese character for "tazuna", a yoke which is used to keep animals together, and means a bond between people that cannot be separated.

In particular, the Japanese archipelago is an inherently unstable land, and we are hit by typhoons every year, and have experienced many major earthquakes. Each time, our peaceful daily lives have been destroyed, and we have lost many people who are indispensable to us. Japanese people have always had a "view of impermanence" as their outlook on life, believing that there is no such thing as "normal" in this world and that things are fleeting. In the process of destruction, they cherished the bonds among people, and in the process of reconstruction, they improved their knowledge and skills and nurtured the spirit of helping one another.

Furthermore, we should not forget that people from all over the world offered their hands of support after the earthquake. The spirit of helping one another in times of trouble is not unique to the Japanese, but is common to all human beings. So, why did only humans have this kind of heart? As a matter of fact, chimpanzees are also said to give a piece of stick to help another chimpanzee who cannot reach a banana. But the chimpanzees, both the one that helped and the one who was helped do not have any feelings about it, they are not grateful for the help, and they do not return the favor on any occasion.

Recently, traces of Homo sapiens, the ancestor of modern humans, were discovered in the Blombos Cave near Cape Town in the southernmost part of Africa. In fact, a tribe with those genes still lives in the village of Mahamasi in the Kalahari Desert. There, food from hunting is limited, so they follow the basic principle of sharing everything with everyone, whether the harvest is good or bad. If a village suffers from hunger, a nearby village will always help, and the village that is helped will give away a lot of milk the following year, thus establishing a mutually beneficial relationship. Those who would not establish such a relationship were driven out of the village and could no longer survive on their own. In the end, only those humans who were able to establish a relationship of mutual help survived, and it is believed that the spirit of "living together" of modern humans was created in this way.

Health, medical, and welfare professionals are professionals who aim to improve people's health and happiness, and it is important to provide care that is close to the human heart. At Kanagawa University of Human Services, new students are immediately given practical training at a neighboring site in collaboration with four departments. Initially, there were various opinions about sending new students, who had no specialized knowledge or skills, out into the field. However, we believed that it is necessary for professionals who provide care for people to improve their sensitivity as human beings in order to be close to people, and that this can be learned through actual experience. In addition, we wanted the students to learn the necessity of multidisciplinary cooperation by forming a team of four departments.

Many students come into contact for the first time with people suffering from incurable diseases, people who were born with mental or physical disabilities, and people who are still trying to live every day. At first, they do not know how to talk to them, and cannot hold their hands or speak to them. This is because today's young people, who have grown up in nuclear families, do not have the experience of witnessing the moment when a person is born or dies, and they rarely see the reality of a person growing old and falling ill. However, becoming a health, medical, or welfare professional means confronting the reality of life, including aging, illness, and death, which can be said to be the fate of human beings.

After the practical training is completed, a presentation is given in the classroom. Many of the students tear up and talk about how moved they are to have a relationship with the sick and disabled that allows them to talk and touch them. I believe that this sensitivity is the key to sustaining and developing this profession.

11.7 Maintenance and Promotion of Physical and Mental Functionality and Nutrition

In order to extend healthy life expectancy, it is necessary to prevent the onset and exacerbation of diseases. And as people age, avoidance of long-term care becomes more important, and the main focus of frailty prevention. At the heart of frailty prevention is the prevention of undernutrition. Typical diseases include energy and protein deficiency, iron deficiency anemia, and osteoporosis due to calcium deficiency. In particular, the elderly suffer from a low-nutritional state in which both energy and protein are insufficient. This is due to a decrease in the intake of fats and oils, meat, milk and dairy products, and eggs, as the elderly generally eat smaller meals and prefer lighter foods. In the elderly, the body's ability to synthesize protein and recover it is also reduced. When energy intake is insufficient, the breakdown of body fat and muscle increases to compensate, resulting in a loss of body weight and muscle mass. This, combined with the decreased ability of the liver to synthesize protein due to old age, is characterized by a decrease in protein (albumin) in the blood.

To what extent do people lose their physical and mental function due to undernutrition?

The decline in physical and mental functioning observed in clinical practice is generally due to the effects of disease and cannot be said to be purely due to nutrition. In order to observe this in an absolute sense, it would be necessary to conduct an experiment of undernutrition in healthy people, a so-called "starvation experiment" which is ethically impossible. However, in 1944, during the war, the "Minnesota Starvation Experiment" was conducted in the United States. The public was invited to participate in the experiment, and for six months, caloric intake was reduced to 1570 kcal per day (half the normal intake), and exercise was limited to walking 35 km per week.

Table 11.1 Psychological changes observed with undernutrition

Decreased ability to concentrate, pay attention, grasp, and make judgments
Increased mental fatigue
Helplessness and lack of emotion increased
Interest in the opposite sex and decreased sexual desire
Addiction to gum and coffee
Increased mood disturbances, fickleness, and irritability
Depression, hysteria
Impatience and frustration developing into angry outbursts
Increased nervousness and anxiety
Nail biting, smoking
Lack of hygiene
Suicide planning and self-harm
Withdrawal, isolation, lack of humor and fraternity
Shoplifting

As a result of the experiment, the average body weight went from 69 to 52.4 kg (−16.6 kg). In addition, physical changes occurred, such as a decrease in body temperature and pulse rate, decreased physical strength, edema, and decreased vision and hearing. Noteworthy were the mental changes such as decreased concentration and attention, depression, irritability, apathy, and hysteria (Table 11.1). These symptoms are often observed in elderly people. In other words, the decline in mental and physical functions is caused by low nutrition, regardless of disease, and the improvement of nutritional status can maintain and improve the mental and physical functions of the elderly to some extent.

11.8   There May Be an Energetic Element in Food

For a long time, I knew I didn't want to get pancreatitis. I had been told by those who had experienced this disease how painful the onset was. In the middle of July 2019, I was hospitalized for 10 days because of severe pain. I usually drink a glass of beer or so, but I got carried away and drank too much because I was enjoying talking with friends at a dinner party and also because good sake was being served. The diagnosis was "alcoholic acute pancreatitis". I had to drink and eat nothing by mouth for a week.

Through three catheters, anti-inflammatory drugs, antibiotics, and nutritional supplements began to be administered. This is the so-called spaghetti syndrome. Three liters of 5% dextrose infusion were administered per day, so the carbohydrate intake was 150 grams and the energy level was 600 kcal. This amount is about half the basal metabolic rate of a healthy person, and because of the inflammation, energy consumption is high, and even if adequate fluids, vitamins, and electrolytes are administered, there is a significant energy deficit. During this period, the breakdown of body fat and protein increases to compensate for the lack of energy and carbohydrates, so my weight at discharge from the hospital had decreased by 3 kg. However, this is the limit of peripheral intravenous nutrition, and even if it is administered for 24 h, a state of energy deficiency cannot be avoided, and there is no administration of amino acids or essential fatty acids. What is important is the nutritional support during recovery.

To tell the truth, strangely enough, not being able to eat was not as hard as I had imagined when I was healthy. It was partly because the intravenous drip kept my blood sugar up and I didn't feel hungry, but all I had to do was walk three meters to the bathroom to urinate, and the rest of the time I just slept and didn't feel any inconvenience because I didn't feel or think anything. I thought that if I continued like this, anyone could enter a "vegetative state", in the usual phrase.

Perhaps it was because I had been well nourished before, but I recovered remarkably quickly, the inflammatory reaction decreased, and I started eating with a soft diet. A soft diet is an easy-to-digest diet containing lots of water.. As I began to feel the delicious taste of food in my mouth and the desire to eat, I began to feel energized and had the strength to try to live. In other words, I went from plant to animal again. It's not that my nutritional intake had been met, my weight had increased, or my albumin had risen. Just being able to eat and taste through my mouth gave me the energy to live positively.

This experience was a good opportunity for me to think about nutrition and food in the future. Since being discharged from the hospital, I have been grateful for being able to eat normally three times a day, as much as I like.

Chapter 12
Nutrition and Diet in the Future

Abstract We now know the causes and solution for the malnutrition that plagues the world. However, the double burden of malnutrition cannot be eradicated. This is because even if nutritional supplements temporarily solve malnutrition, it is necessary to adjust the daily diet to maintain good nutritional status continuously, universally, and for a lifetime. This means that malnutrition can be eradicated only if all people have access to a sustainable healthy diet. It is not easy to create a situation where everyone has access to such a diet, because the daily diet depends on a industrial system centered on food, and this system is influenced by a wide variety of factors.

The food that makes up a diets depends on a system of production, processing, distribution, cooking, and presentation, and each of these processes is influenced by the food environment and food culture. If we consider the human side, the human body consists of cells and organs, and in addition humans also depend on mental and social aspects of life within the framework of family, community, and nation.

Traditionally, nutrition science has examined the relationships among the factors that impact food and humans. These conventional research has clarified the relationship between each factor such as meal and organs and diseases, food and nation, and menu and community etc. However, only these methods do not lead to a way to make sustainable healthy eating accessible to all. Both the food on the left axis and the people on the right axis are organisms that exist on the same finite planet, and if the earth loses its ability to sustain them, this system will collapse from the foundation up.

In order to achieve this, it is necessary for all organizations and people involved in nutrition and diet to raise a unified banner of "eradicating malnutrition" and to strive to set feasible goals in their respective areas. The comprehensive measures are needed through collaboration among governments, universities, industry, economy, finance, etc.

Keywords Decade of action for nutrition · Tokyo nutrition summit 2021 · The double burden of malnutrition · Nutritional structure · Eradicating malnutrition · Industry-government-academia collaboration · Science technology innovation basic law · Kanagawa University of Human Services School of Health Innovation · COVID-19 and nutrition · Green recovery · International Energy Agency (IEA) ·

Livestock's Long Shadow (FAO) · Planetary healthy diet · Japan's National Institute for Global Environmental Strategies · 1.5-Degree Lifestyles: Targets and Options for Reducing Lifestyle Carbon Footprints · Tokyo Nutrition for growth Summit 2021 · Universal health coverage (UHC)

12.1 Why Hasn't Malnutrition Disappeared

The United Nations has formulated the Decade of Action for Nutrition (2016–2025) and the Sustainable Development Goals (SDGs). Furthermore, The Japanese government have announced that a "Tokyo Nutrition Summit 2021" will be held in December 2021. Now is the perfect opportunity to put an end to global malnutrition. However, "2018 Global Nutrition" reports with a sense of urgency that "the reality of nutrition is unacceptably poor and progress in improving it is not being made".

This is because there has been no improvement in eight key nutritional indicators: hypertension, obesity and overweight in adults; stunting, wasting and overweight in children; and anemia and excess salt intake at all ages. 22.2% (150.8 million) of the world's children are stunted, 7.5% (50.5 million) suffer wasting, and 5.6% (38.3 million) are overweight.

In recent years, nutritional science has made remarkable progress, about 35–40 different nutrients have been discovered, and their functions, foods containing them, and nutrient deficiencies and excesses have been identified. The development of comprehensive nutritional foods and nutritional supplements that contain all the nutrients has made it possible to correct malnutrition and to administer all the nutrients directly into the digestive tract and bloodstream through a tube or drips, even in cases where eating is difficult due to illness or disability. In other words, we now know the causes and solution for the malnutrition that plagues the world.

However, the double burden of malnutrition cannot be eradicated.

This is because even if nutrition and nutritional supplements temporarily solve malnutrition, it is necessary to adjust the daily diet to maintain good nutritional status continuously, universally, and for a lifetime. This means that malnutrition can be eradicated only if all people have access to a sustainable healthy diet. On the other hand, it is not easy to create a situation where everyone has access to such a diet, because the diet depends on a huge industrial system centered on food, and this system is influenced by a wide variety of factors.

For example, the food that makes up a meal depends on a system of production, processing, distribution, cooking, and presentation, and each of these processes is influenced by the food environment and food culture.

Even if a rural life of gathering vegetables from the garden and keeping chickens for eggs and meat is partially achieved through home gardens, it is not possible to feed the global population in this way. The vast majority of people live on food produced and distributed by giant agricultural and food corporations, and it will take a comprehensive force to make this system work properly.

If we consider the human side and the energy and nutrients required for sustaining life, the human body consists of cells and organs, and in addition humans also depend on mental and social aspects of life within the framework of family, community, and nation. Traditionally, nutrition science has examined the relationships among the factors that impact food and living things (humans), beginning with energy and nutrients on the left of the chart of the chart below and extending to food, cooking, menus, health and supplements, and even to the role of the state on the right. However, these methods, even if they increase the integration' of the left-right relationships, reveal only the connections between various parts of the whole , and do not lead to a way to make sustainable healthy eating accessible to all.

Moreover, both the food on the left axis and the people on the right axis are organisms that exist on the same finite planet, and if the earth loses its ability to sustain them, this system will collapse from the foundation up.

To sustain life and develop a healthy and cultured life for the future, without leaving anyone behind, we must create a world where everyone has access to a healthy diet that recognizes biodiversity and does not burden the global environment. In order to achieve this, it is necessary for all organizations and people involved in nutrition and diet to raise a unified banner of "eradicating malnutrition" and to strive to set feasible goals in their respective areas. This is why comprehensive measures are needed through collaboration among governments, universities, industry, economy, finance, etc.

12.2 Integrated Nutrition and Innovation

I have already mentioned the need for individuals and organizations involved in nutrition to work together to increase their collective effectiveness in order to eradicate malnutrition, and in fact, The science of nutrition in Japan has been putting this into practice.

The improvement of nutrition in Japan started 150 years ago, when the nation was modernized by the Meiji Restoration, and nutrition as a science was introduced. In particular, it was around 1890, the time of the Sino-Japanese War and the Russo-Japanese War, that the nation began to make serious efforts. The government addressed the issue of nutrition as a national policy in order to raise healthy citizens and strong soldiers under the national policy of "Wealthy Nation, Strong Army". Nutritional deficiencies due to food shortages were particularly serious before and after World War II, and the country was plagued by severe malnutrition and a variety of nutritional deficiency diseases. The government imported food from the United States and at the same time imposed food controls. It also educated people about the importance of nutrition and how to improve nutrition in order to consume the limited food rationally. For this purpose, the government trained nutritionists whose professional mission was to improve nutrition and established a national qualification for them based on the Status Law. Moreover, the "Nutrition Improvement Act" was enacted so that nutritionists could be placed in every corner of society.

Nutritionists were assigned not only to hospitals and schools, but also to day-care centers for infants, kindergartens, welfare facilities, enterprises, the Self-Defense

Forces, prisons, and all other facilities where meals were served. In 1954, the "Japan Society for the Improvement of Nutrition" was established to study methods of improving nutrition, and improvement of nutrition became the theme of scientific research.

In recent years, when I meet with representatives of countries suffering from malnutrition, I ask them, "Are there any nutrition experts in your country?" The representative of the country introduce me to a well-known nutritionist, who shows me the nutritional problems of the country and explains at length the results of his or her research on them. I then ask the following question.

"I understand that your research deserves a Nobel Prize. By the way, what are you going to do about those skinny children and those old men whose belts look like they are about to fall down that you see so often in this town?"

In many cases, I don't get a decent answer.

There are so-called nutrition researchers all over the world, and the more serious the nutrition problem is in a country, the more famous the nutritionists are. By the same token, the country with the world's best obesity research has the world's highest obesity rates.. What I would like to know is whether or not there are nutritionists, or nutrition professionals, who can provide practical guidance on improving nutrition. In many countries, the training of nutritionists is inadequate, or their work is limited to medical treatment and they are positioned as leaders in dietary therapy. In Japan, more dietitians and nutritionists are trained than in any other country in the world, and they are placed not only in medical care but also in health and welfare facilities, contributing to health promotion and disease prevention.

Furthermore, one of the characteristics of nutrition in Japan is the contribution of the food industry. This is because the dawn of the modernization of the food industry in Japan occurred in many cases with the aim of improving the nutrition of the people. For example, Ajinomoto Co., Ltd developed a seasoning made from the umami ingredient (glutamic acid) discovered in kelp "dashi" by Dr. Kikunae Ikeda of Tokyo Imperial University, and popularized it as an easy and tasty way to make simple meals more palatable. Subsequent research led to the addition of umami to the four primary flavors. The founder of Ezaki Glico Co Ltd., Riichi Ezaki, collected glycogen from oyster broth and added it to caramel, and the founder of Yakult Honsha Co., Ltd Dr. Minoru Shirota, popularized "lactobacillus beverages" as a way to "contribute to the health of the people". Dairy manufacturers such as Meiji, Morinaga, and Snow Brand worked to popularize milk and dairy products as foods that provided high-quality protein, vitamins, and minerals, which were in short supply at the time. School lunches, which are highly regarded internationally as a model of nutrition education for children, also began with milk served during lunch period after the war. The Takeda Pharmaceutical Company Lid fortified white rice with vitamin B_1 to prevent beriberi, and Otsuka Pharmaceutical and Taisho Pharmaceutica popularized many kinds of nutritional foods. Oil and fat manufacturers such as Nisshin Oillio contributed to the westernization of meals by popularizing seasoning oil. Furthermore,

Kagome Co., Ltd. produced ketchup, an essential ingredient in Western-style meals, and House Foods and S&B Foods developed the technology to produce curry in Japan, which had been adopted by the British Navy as a highly nutritious and delicious dish. The Japanese food industry had a strong desire on the part of its founders to free people from poverty and make their diet richer and healthier.

Because Japan's food processing technology and distribution system had not developed as much as those of advanced Western countries, local cuisine, which was mainly rice and items processed from locally produced ingredients, flourished and became traditional. In other words, in order to modernize the nation, the movement to improve nutrition was based on nutritional science. However, in the natural environment in each region, people did not oppose nature, but enjoyed the changes of the four seasons, devised cooking and preservation methods, and utilized seasonings to create delicious and healthy meals.

"Japan Nutrition" was a major project that demonstrated the comprehensive power of the "industry-government-academia collaboration" that has been called for in recent years. Such a comprehensive approach to nutrition was effective in resolving postwar nutritional deficiencies, and at the same time, because many citizens learned the importance of nutrition and the basics of healthy eating, it was also effective in preventing and eliminating obesity and metabolic syndrome, which emerged after Japan's rapid economic growth.

The Japanese government changed the existing " Science and Technology Basic Law" to the " Science Technology Innovation Basic Law" starting in FY2021. The goal is not only to develop science and technology, but also to encourage research to implement the results, and ultimately to create innovation. The innovation is not limited to product development, but includes a broad range of activities that bring about major changes in society, creating new value and transforming society itself.

In April 2019 Kanagawa University of Human Services established the School of Health Innovation (SHI). Focusing on public health, the school aims to bring innovation to health, medicine, and welfare. In other words, the School seeks to create the next generation of health innovators, and as an affiliated institution of the entire university, it is an organization that performs think tank functions in order to support policy making and promote academic research and its social implementation. In addition to conducting policy research and making recommendations on health, medical care, and welfare, the institute promotes the establishment of innovative pre-disease and healthcare industries, and matches academic research with real-world issues to facilitate the social implementation of innovation.

We hope that in the future, all countries will be able to eliminate malnutrition and create nutritional and dietary innovation that is kind to the planet and to people, and to enable all people to live healthy and happy lives.

12.3   COVID-19 and Nutrition

In November 2019, a new type of coronavirus emerged, and by late January of the following year, it had spread to mainland China, followed by East Asia, Europe, and the rest of the world. On January 31, 2020, the WHO declared it a "public health emergency of international concern" and urged people around the world to take precautions. As of March 7, the spread of the disease had not stopped, with the number of infected people exceeding 100,000 worldwide, and on March 11, the WHO recognized that the spread of the disease was equivalent to a pandemic. The term "pandemic" is derived from the Greek word.

On April 7, 2020, amidst a flurry of policy announcements by the Japanese government, Prime Minister Abe declared a state of emergency because of the new coronavirus infection. All countries appealed to their citizens to stay at home and not to go out, and people disappeared from towns and cities, and were banned from eating out, going to school, traveling, and going to work. However, mankind has historically gone through this many times and survived, each time advancing science and culture and creating new social structures. The problem is that in this battle, many casualties occur, and the brakes are put on the activities of science, art, culture, etc. and it becomes impossible for some fields to function and they disappear.

There are hundreds of thousands of unknown viruses on the earth, and with the thawing of the frozen soil due to global warming and the development tropical forest areas, the opportunities for human contact with viruses are increasing. Each time a new virus emerges , people's lives are affected, and it takes time for a vaccine to be developed. During an outbreak, in order to reduce the number of victims as much as possible, it is necessary to ensure good hygiene habits and to build up the body's resistance. In order to enhance the body's protective functions, it is essential to maintain good nutritional status. The Japan Dietetic Association sent out a message ahead of the rest of the world on April 3, 2020. The purpose of this message was to say, "Since many nutrients are involved in maintaining immunity, it is necessary to maintain a healthy diet even when daily life is restricted" (Table 12.1).

In addition COVID-19 has been found to be exacerbated by obesity associated with overnutrition. This is because the production of anti-inflammatory adiponectin from visceral adipocytes is reduced, leading to chronic inflammation of adipose tissue, and the disease is exacerbated by a cytokine storm, which causes the immune system to run amok when infected with a [the?] virus. Belanger and colleagues have shown that not only is obesity involved in the exacerbation of COVID -19, but obesity also puts people at risk for chronic diseases such as diabetes, and that these diseases are also factors in the exacerbation of COVID -19. Ultimately, access to a healthy diet is important for the prevention and moderation of COVID -19.

Table 12.1 Overcoming difficulties with the power of nutrition

On Tuesday, April 7 in 2020, a state of emergency was declared by the Headquarters for the COVID-19. I would like to express my gratitude and sincere respect to all of the dietitians, nutritionists, and medical and welfare professionals who continue to work at the front lines of the situation. I am also concerned about our members who are working in a variety of workplaces under conditions that they have never experienced before. With the spread of the new coronavirus, it is becoming increasingly difficult to maintain a daily diet. In light of this situation, I would like to send out a message to you on behalf of the Japan Dietetic Association. To prevent viral infections, it is recommended to avoid close contact as well as to wash hands thoroughly and to wear a mask. In addition to these precautions, there is one more thing to be careful about. This is to maintain and strengthen your "immunity" against the virus.

Recent studies have shown that our immunity is based on the complex metabolism of diverse components, and since many types of nutrients are involved in this mechanism in various ways, malnutrition reduces immunocompetence. A typical form of malnutrition is Protein-Energy Malnutrition] (PEM). In the elderly, emaciation and lowered serum albumin levels have been found to significantly reduce both the antibody-positive rate after vaccination and the rate of infection prevention. In addition, since various vitamins act as coenzymes for various metabolisms, deficiency of these vitamins leads to a decrease in the cellular functions of immunity. Mineral deficiencies can lead to thymus dysplasia and reduced levels of immunoglobulins, which are antibodies.

Currently, immune-related nutrients include energy, protein, n-3 fatty acids, dietary fiber, vitamin A, vitamin D, vitamin E, vitamins B (B_1, B_2, B_6, B_{12}, folic acid, pantothenic acid, niacin, and biotin), vitamin C, iron, zinc, copper, and selenium. Lactobacillus acidophilus and lactoferrin are also being considered as related ingredients. In other words, since many nutrients are involved in immune function in a comprehensive manner, it is important to continue to eat a healthy diet that provides all nutrients without excess or deficiency, rather than relying on one particular nutrient or food.

Voluntary restraint in going out, the decline in food production and distribution capacity, and consumers buying in bulk have led to a bias toward the purchase and consumption of certain foods, making it difficult to have a nutritionally balanced diet. Now is the time to utilize the power of nutrition and have the strength to overcome the COVID-19.

Dietitians and nutritionists across the country should join forces with the public and do our utmost to overcome this difficult situation. And we sincerely hope that all our members and their families will stay safe.

Japan Dietetic Association, President: Teiji Nakamura

12.4 "Green Recovery" and Future Nutrition

The International Energy Agency (IEA) has reported a significant reduction in global greenhouse gas emissions as a result of COVID-19's urban closings and travel restrictions. Ironically, COVID-19 reduced the burden on the global environment: in April 2020, the environment ministers of the EU and other countries proposed "Green Recovery" as a recovery plan for COVID-19. Green Recovery is a plan that aims to develop the post-COVID-19 reconstruction of the economy while reducing the emission of greenhouse gases that cause environmental impact.

What role should nutrition and diet play in a green recovery plan?

According to the FAO report "Livestock's Long Shadow", greenhouse gas emissions from transportation and livestock farming account for 13.5% and 18% of total emissions, respectively, indicating that meat-eating has a particularly large environmental impact. Livestock feed, manure and even burps increase the amount of emissions. In January 2019, the Lancet published "Food for the Anthropocene: The EAT Lancet Commission on Healthy Eating through Sustainable Food Systems", which stated that meat consumption should be reduced to 14 g per day. This philosophy is the basis of the proposed "Planetary Healthy Diet", in which half the plate consists of fruits, vegetables, and nuts, and the other half of whole grains, vegetable proteins, unsaturated vegetable oils, moderate amounts of meat and dairy products, and sugar and starchy vegetables.

In February 2019, Japan's National Institute for Global Environmental Strategies, along with others, published "1.5-Degree Lifestyles: Targets and Options for Reducing Lifestyle Carbon Footprints". The total amount of greenhouse gases emitted by an average Japanese person in a year is 7.6 tons per person. Of this, 2.4 tons are from housing and electricity, 1.6 tons from transportation, cars, etc., and 1.4 tons from food. Of the dietary sources, meat and milk and dairy products combined account for 0.5 ton, or 6.6% of the total emissions of Japanese people. This value is less than half of the global average of 14.5% as determined by FAO (Fig. 12.1).

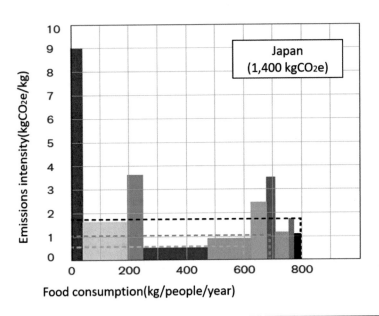

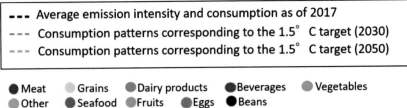

Fig. 12.1 Breakdown of Japanese carbon footprint by food group (2017)

Per capita GHG emissions from current diet

Per capita GHG emissions from dietary guidelines

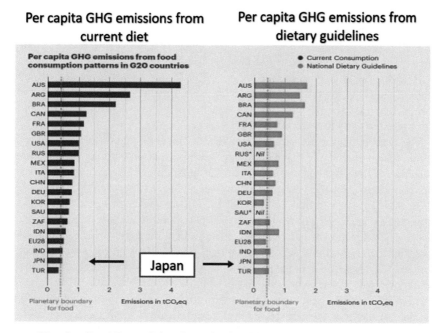

Diets for a Better Future: Rebooting and Reimagining Healthy and Sustainable Food Systems in the G20

Fig. 12.2 Japanese food has a low environmental impact and is similar to planetary diets 'kg/person/year'

In other words, Japanese people consume protein from seafood and soybean products, and their consumption of meat, milk, dairy products, and other livestock is consequently low, so there is no need to restrict livestock as drastically as in the West. The Lancet Commission analyzed the diets of 20 industrialized countries from the viewpoint of environmental impact, and reported that in the case of the Japanese, both the current diet and the content of the dietary guidelines are the same as the ideal Planetary Healthy Diet (Fig. 12.2).

12.5 Promoting Japan Nutrition at the "Tokyo Nutrition for Growth Summit 2021"

The Japanese government has announced that the Tokyo Nutrition Summit 2021 will be held in December. The summit will be hosted by the Japanese government and co-hosted by the World Health Organization (WHO), the Food and Agriculture Organization (FAO), the World Food Programme (WFP), the United Nations

Children's Fund (UNICEF), and the World Bank ; leaders and ministers from various countries are expected to attend.

The "Nutrition Summit" is an international effort to improve nutrition initiated by the United Kingdom and formed of the London Olympics.

Held for the first time in London in 2013, it was triggered by the "Hunger Summit" held by then British Prime Minister Cameron. The second time at the Rio de Janeiro 2016 Olympic Games, and it will be carried over as the "**Tokyo Nutrition for growth Summit 2021**"

The main themes of the summit are: Health: integrating nutrition into universal health coverage (UHC); Food: building healthy and sustainable food systems; Resilience: combating malnutrition in fragile situations; Accountability: data-based monitoring; and Finance: securing financial resources to improve nutrition.

At the "Tokyo Nutrition Summit 2021", Japan's approach to nutrition will be presented as a package. The Ministry of Health, Labor and Welfare (MHLW) aims to visualize the direction of Japan's efforts in nutrition policy and to make international contributions in nutrition. Japan is a country that has tackled head-on the double burden of under-nutrition in the prewar and postwar periods and over-nutrition in the period of rapid economic growth, and has built a nation of longevity. The presentation will introduce the nutrition policy from the post-war reconstruction period to the present day, the nutrition policy in the context of an aging society with a low birthrate, disaster prevention nutrition that supports the life and livelihood of residents even in times of disaster, the training and activities of professionals such as dietitians, nutritionists, dietary life improvement promoters, and nutrition teachers, as well as the efforts of industry-government-academia collaboration. The Japan Dietetic Association plans to make a commitment to Asian and African countries suffering from malnutrition by providing assistance in the education and training of dietitians, who are key players in the eradication of malnutrition, and in the development of dietitian systems.

At present, the system to eradicate malnutrition world-wide is better prepared than ever before and the situation is gradually improving. I would like to call on this international conference to accelerate the speed of this process and the need for further expansion. It is significant that the Nutrition Summit is being held in Japan at such a time. While inheriting a traditional food culture that respects nature, Japan introduced nutritional science after the Meiji era (1868–1912), and solved malnutrition through comprehensive nutritional improvement. Even after its rapid economic growth, Japan worked on the prevention of obesity and non-communicable diseases, and formed a nation with a long and healthy life. We hope that Japan can contribute to the health and well-being of people around the world by introducing to the world this unique Japanese nutritional improvement that combines "culture and science"

I collectively called these things "Japan Nutrition" and decided to disseminate them to the world (Table 12.2).

Table 12.2 What is "Japan Nutrition"?

①	In the era of staple foods and frugal diets, Japanese people were deficient in protein, essential fatty acids, and various vitamins and minerals, and suffered from many nutritional deficiencies.
②	Many people died of tuberculosis and other infectious diseases because of low nutrition, high infant mortality, and lack of resistance.
③	The combination of low nutrition and excessive salt intake led to high blood pressure, strokes, and even death from stomach cancer, and the Japanese had a short life expectancy.
④	Comprehensive nutritional improvement based on nutrition education through industry-government-academia collaboration helped to solve [to remove/alleviate] the double burden of malnutrition.
⑤	Inheriting the Japanese food culture that respects nature, people created a sustainable and healthy diet that leaves no one behind.

Bibliography

Belanger MJ et al (2020) Covid-19 and disparities in nutrition and obesity. https://doi.org/10.1056/NEJM2021264

Diets for a Better Future: Rebooting and Reimagining Healthy and Sustainable Food Systems in the G20

FAO (2006) Livestock's Long Shadow environmental issues and options

FAO (2013) Tackling climate change through livestock- a global assessment of emissions and mitigation opportunities

National Dietetic Association (ed) (1981) Dietitian act and nutrition improvement act. In: History of dietetics in Japan, pp 252–258, Group Meal by Release Food, pp 213–226, Gakken Medical Shujunsha Co., Ltd.

Report by the National Institute for Global Environmental Strategies and others, 1.5-Degree Lifestyles: Targets and Options for Reducing Lifestyle Carbon Footprints, 2019

Shiro Yakuwa (2009) Trends in the dietitian system. Journal of the 50th Anniversary of the Incorporation of Japan Dietetic Association, 26–57, Japan Dietetic Association

Toshio Oiso (1980) At last, the nutrition administration regains its vitality. In: Insatiability in the midst of confusion, Ishiyaku Publishers, Inc., pp 197–224

Willet W et al; Food in the Anthropocene: the EAT-Lancet Commission on healthy diets from sustainable food systems, Lancet, Published online January 16, 2019 https://doi.org/10.1016/S0140-6736(18)31788-4

World Health Organization, Food and Agriculture Organization of the United Nations, and WHO, "Sustainable Healthy Diets Guiding Principles," 2019

World Nutrition Report: "Global Nutrition Report 2014" International Food Policy Research Institute 2014

World Nutrition Report: "Global Nutrition Report 2018" International Food Policy Research Institute 2018

Printed in the United States
by Baker & Taylor Publisher Services